Österreichische Akademie der Wissenschaften

Mathematisch-naturwissenschaftliche Klasse

Anzeiger

Abteilung I

Biologische Wissenschaften und Erdwissenschaften

138. Band
Jahrgang 2007

T0132858

Wien 2008

Verlag der Österreichischen Akademie der Wissenschaften

Inhalt

Anzeiger Abt. I

Anzeiger Abt. I (2007) 138: 3–15

Anzeiger
Mathematisch-naturwissenschaftliche Klasse Abt. I
Biologische Wissenschaften und Erdwissenschaften

Über die wissenschaftlichen Beziehungen von Hans Ertel und Heinrich Ficker

Von

Wilfried Schröder

(Vorgelegt in der Sitzung der math.-nat. Klasse am 15. November 2007
durch das w. M. Helmut Moritz)

1. Vorbemerkung

Im Gegensatz zum englischen Sprachraum befindet sich die Aufarbeitung der Geschichte der geophysikalischen Disziplinen in Deutschland erst im Aufbau. Das ist bedauerlich, denn gerade hier wurden viele Grundlagen der heutigen Meteorologie und Geophysik gelegt. So kann es keinen Zweifel geben, dass gerade auch das Berliner Meteorologische Institut zur Zeit der Wirksamkeit von HEINRICH [VON][1] FICKER sowie HANS ERTEL eine internationale Leitfunktion hatte. Die Wiederkehr des 50. Todesjahres von Akademiemitglied HEINRICH FICKER mag Anlass sein, einige Linien in der Geschichte der Beziehung dieses Wissenschaftlers zu ziehen. Vornehmlich wird man sich dabei auf HANS ERTEL beziehen, dessen Entdeckung FICKER selbst als seine wohl bedeutendste wissenschaftliche Leistung bezeichnete. Von einem Versuch muss indes gesprochen werden, weil viele Unterlagen aus der Zeit der beiden Gelehrten nicht mehr verfügbar sind, so dass auf manche Linie indirekt geschlossen werden muss.

---

[1] Je nach dem verschiedenen Gebrauch der ehemaligen Adelsprädikate in Deutschland und Österreich.

2. Die Berliner Zeit

Die meteorologische Forschung hat im Berliner Raum eine lange Tradition. Die Gründung des „Preußischen Meteorologischen Instituts" im Jahre 1847 bildet dabei in organisatorischer Hinsicht nur den Beginn.

2.1. Entwicklungslinien

Auf WILHELM MAHLMANN (1812–1848) folgte HEINRICH WILHELM DOVE (1803–1879), der die meteorologische Arbeit in vielfältiger Weise aktivieren sollte. Eine Neuorientierung des Instituts erfolgte dann durch den Physiker WILHELM VON BEZOLD (1837–1907). In den Folgejahren gewann auch die theoretische orientierte Forschung zunehmend Gestalt, wobei der Name von HERMANN VON HELMHOLTZ (1821–1894) als Beispiel genannt sei. So wurde auch rasch die Bedeutung der Helmholtzschen Wirbelsätze für die Meteorologie gesehen. DOVE und BEZOLD waren es, die die Wirbeldynamik auf die reale Atmosphäre zu übertragen suchten. Erst in unserer Zeit hat indes HANS ERTEL (1904–1971) ihr jene Form gegeben, die auch für reale Medien anwendbar ist.

Im Jahre 1923 übernahm HEINRICH FICKER als Nachfolger von GUSTAV HELLMANN (1854–1939) das Amt sowie den Lehrstuhl in Berlin. Bereits die Antrittsrede FICKERS vor der Preußischen Akademie lässt einige Aspekte seines Programms, wenn man es so bezeichnen will, erkennen. FICKER war am 28. Juli 1926 zum Ordentlichen Mitglied geworden und hielt am 30. Juni 1927 seine Antrittsrede. Darin hebt er sowohl die Beiträge der österreichischen als auch der skandinavischen Meteorologenschaft hervor. Besonders die klassischen Studien von HANN, EXNER, MARGULES und TRABERT betont er. Interessant ist auch, dass FICKER gerade in den MARGULESSchen und EXNERSchen Beiträgen wesentliche Vorarbeiten für die später formulierte Polarfronttheorie von BJERKNES sieht. Damit ist etwas angesprochen, das über Jahre hinweg die Gegensätze zwischen der norwegischen Schule einerseits, und der österreichisch-deutschen Schule andererseits, beinhaltet. Dies ist in gewisser Weise fast tragisch zu nennen, denn der Beginn der wissenschaftlichen Beziehungen zwischen BJERKNES und FICKER war überaus freundlich und von aufrichtigem, gegenseitigem Respekt geprägt. Einige Aspekte aus diesen Beziehungen müssen angesprochen werden, weil sie zum Verständnis der späteren Jahre wichtig sind.

2.2. Bjerknes und Ficker

Das Problem der „Polarfronttheorie" kommt einerseits in einem Brief FICKERS an BJERKNES vom 15. Oktober 1920 zum Ausdruck. Darin betont er, dass er anlässlich eines Leipziger Besuches mit diesem das Problem erörtern konnte. FICKER beklagt, dass – infolge der Kriegsauswirkungen – seine Arbeiten zu wenig Beachtung gefunden haben. In seiner kritischen Lage bittet er BJERKNES deshalb sowohl um allgemeine Literatur als auch selbst nur solche zur Polarfronttheorie. Unter dem 30. Oktober 1920 schrieb BJERKNES sofort zurück und bemerkte:

„... Leider können wir deshalb nur sehr wenig zurückschicken. Denn seitdem wir den „praktischen Weg" eingeschlagen haben, hat uns die „Praxis" überwältigt. Im Wesentlichen kann ich Ihnen daher nur ältere Arbeiten schicken, die Ihnen zeigen werden, von welchem abstrakt entlegenen Gebiete ich in die Meteorologie hineingeraten bin. Unwissenheit in der meteorologischen Literatur müssen Sie deshalb auch mir und den Meinigen vergeben. Hier in Bergen sind wir außerdem oder waren wenigstens zu Anfang so ziemlich isoliert."

Unter dem 12. November 1920 bemerkt FICKER seine Freude über BJERKNES' Hinwendung zur Praxis:

„Dass Sie, hochverehrter Herr Professor, von so abstrakt-theoretischer Seite in die Meteorologie gekommen sind, ist uns ein ungeheuerer Vorteil auch für die meteorologische Praxis. Der Praktiker sieht ja so oft den Wald vor lauter Bäumen nicht."

Zwei Jahre später macht FICKER in einem Brief an BJERKNES indes bereits auf unterschiedliche Standpunkte aufmerksam, wenn er schreibt:

„Was das Depressionsschema selbst betrifft, so habe ich bezüglich einiger Einzelheiten allerdings eine abweichende Auffassung. Nach den Beobachtungen liegt die Stratosphäre nicht am tiefsten über dem warmen Gebiete, sondern erreicht ihre tiefste Lage erst dann, wenn in den niederen Schichten bereits eine niedrige Kältewelle im Gange ist. Auch bezüglich der Vorgänge an der warmen Front bin ich zum Teil anderer Ansicht. Ich vermisse in Ihrem Depressionsschema z. B. das antizyklonale Entwicklungsstadium mit absteigender Luftbewegung ... . Der Hauptunterschied unserer Auffassungen liegt wohl darin, dass nach meiner Meinung die Vorgänge in der Stratosphäre eine gewisse, aktivere Rolle spielen als ihnen in Ihrer Auffassung eingeräumt wird." (FICKER, 27. Januar 1922.)

Das Thema „Rolle der Stratosphäre für das Wettergeschehen" sollte weiterhin die Beziehungen zwischen BJERKNES und FICKER bestimmen. Es führte auch zu einer gewissen Entfremdung der beiden Schulen. Es wundert nicht, dass zunächst FICKER sowie später ERTEL dieses Thema erneut aufgriffen (siehe ERTEL [3]). Die Unterschiede zwischen BJERKNES und FICKER kommen in einem weiteren Schreiben FICKERS vom 27. Februar 1923 zum Ausdruck:

„... Ich habe über Ersuchen EXNER's für die Meteorologische Zeitschrift ein langes Referat über die Polarfronttheorie geschrieben. Ich erlaube mir, seine

Korrektur an Sie nach Gastein zu schicken. Dass ich sachlich auf dem Boden der Polarfronttheorie stehe, ist Ihnen ja wohl bekannt. Ich fühle mich aber andererseits verpflichtet, über die ähnlich gerichteten Bestrebungen der österreichischen Meteorologie zu referieren, andererseits habe ich die Empfindung, dass die neuesten Ergebnisse der Theorie-Periodizität der Familien etc. viel mehr als bisher durch empirisches Beweismaterial gestützt werden müssen. Es wäre für mich, obwohl ich ausgesprochener und einseitiger Empiriker bin, natürlich von höchstem Werte, wenn mir Ihr Besuch in Graz Gelegenheit zu einer persönlichen Aussprache mit dem Schöpfer der Polarfronttheorie geben würde."

Der weitere, nur sporadisch vorliegende Briefwechsel betrifft die Vortragsreise von BJERKNES nach Berlin. In einem weiteren Schreiben beglückwünscht FICKER BJERKNES zu dessen Wahl zum Korrespondierenden Mitglied der Berliner Akademie der Wissenschaften (10. Mai 1928).

2.3. Das Berliner Meteorologische Institut

Im Jahre 1923 wurde FICKER Professor für Meteorologie an der Universität Berlin sowie Direktor des Preußischen Meteorologischen Instituts. Über seine Absichten schreibt er an BJERKNES noch am 27. Februar 1923:

„Es wäre für mich von umso größerem Werte, als ich in meiner neuen Stellung als Direktor des Preußischen Meteorologischen Institutes meine Hauptaufgabe darin erblicken werde, die wissenschaftliche Tätigkeit des Institutes in modernere, von Ihnen vertretene Bahnen zu lenken."

In späteren Jahren ergaben sich indes Änderungen in der Organisation des Instituts, das anfangs auch durchaus noch für die Lehrzwecke genutzt wurde. Der größte Teil wurde jedoch 1934 in den „Reichswetterdienst" überführt. Das was übrig blieb, bildete das Universitätsinstitut, das FICKER als Direktor leitete (neben seiner Lehrtätigkeit an der Universität sowie an der Akademie der Wissenschaften). Hinzu kommt, dass er von 1928 bis 1945 zugleich Präsident der Internationalen Klimatologischen Kommission war. Vor der erfolgten Trennung hatte die meteorologische Bibliothek Präsenzcharakter, d. h. die vorhandene Literatur konnte an Ort und Stelle eingesehen und genutzt werden. Die Verlagerung zum Reichswetterdienst brachte zwar diesem einen Aufwärtstrend, machte jedoch für das verbliebene Universitäts-Institut neue Überlegungen notwendig. Neben einem Neuaufbau bezüglich der Literatur wurde schließlich eine eigene Schriftenreihe begründet, die FICKER mit ERTEL herausgeben sollte.

Für die Berliner Zeit war neben FICKER besonders A. DEFANT sowie J. BARTELS wichtig. BARTELS war seit 1927 Privatdozent an der Universität und wurde 1928 Professor. 1934 wurde er Professor für Geophysik an der Forstwirtschaftlichen Hochschule in Eberswalde, 1936 war er Professor für Geophysik an der Berliner Universität und übernahm zugleich die Leitung des Geophysikalischen Instituts in Potsdam. BARTELS war zusammen mit FICKER Referent der ERTELschen Dissertation „Theorie der durch Variationen des magnetischen Potentials induzierten Erdströme bei ungleichförmiger Leitfähigkeit der Erdrinde", mit der ERTEL mit Auszeichnung promovierte. Vorübergehend wirkte ERTEL am Meteorologisch-Magnetischen Observatorium in Potsdam, doch konnte es ihn dort nicht lange halten. Zwar hätte es BARTELS sehr gerne gesehen, wenn ERTEL bei ihm geblieben wäre, aber diesen zog es zurück nach Berlin.

Neben der bereits von FICKER beabsichtigten modernen Ausrichtung der Forschung, der auch das spätere Universitäts-Institut galt, wurden auch weitere Teilfragen der Meteorologie umfassend gepflegt. Die Wettervorhersage verblieb dem Institut: FICKER selbst kam täglich zur Mittagsprognose. Arbeiten zum Bio-, Agrar- und zum Lokalklima wurden gefördert. Die theoretische Ausrichtung lag später bei HANS ERTEL, der zudem ab 1938 als apl. Privatdozent die Theoretische Meteorologie vertrat. Übrigens fallen in diese Zeit wichtige Begegnungen. So weilte der bekannte spanische Meteorologe FRANCESCO MORAN-SAMANIEGO in Berlin, machte ERTELS Bekanntschaft, die jahrzehntelang dauerte. Der international bekannte amerikanische Meteorologe JEROME NAMIAS hatte seinerzeit beabsichtigt, in Berlin zu studieren und mit ERTEL zusammenzuarbeiten. In gewisser Weise hatte das frühere Universitäts-Institut vor über 70 Jahren eine sehr nachhaltige internationale Ausstrahlung gehabt, die auch zunehmend mit dem Bekanntheitsgrad von HANS ERTEL zusammenhing, dessen Arbeiten auch in England und USA sehr wohl zur Kenntnis genommen wurden.

Es muss ja auch gesehen werden, dass in der internationalen Wissenschaft Deutsch die Fachsprache war und die „Meteorologische Zeitschrift" eine international führende Zeitschrift gewesen ist. Fast alle international bekannten Meteorologen haben darin publiziert.

Bereits frühzeitig hatte FICKER ERTELS Begabung erkannt, und so versuchte er, ihn an die Meteorologie zu binden. Andererseits hätte auch BARTELS gerne ERTEL an seinem Institut gehabt. So kam es, dass ERTEL kurzfristig in Potsdam arbeitete, jedoch schnell in die Meteorologie zurückkehrte. Aus dieser Zeit stammen auch einige Arbeiten aus dem Interessensgebiet BARTELS, wie z. B. zur Grönland-

drift, Bewegung von Elektronen in inhomogenen Magnetfeldern
sowie zur Polfluchtkraft, Themen, die ERTEL aufgegriffen hat und die
auch später in der Literatur immer wieder zitiert wurden.

ERTEL legte FICKER eine Arbeit vor, von der FICKER sagte, er habe
sie zwar nicht verstanden; gleichwohl ließ er sie drucken. ERTEL hatte
somit mit 25 Jahren seine erste wissenschaftliche Publikation herausgebracht. Nachdem FICKER auf ERTEL aufmerksam geworden
war, tat er alles, um diesem den akademischen Weg zu ebnen. In
kurzer Zeit absolvierte ERTEL an der Universität in Berlin sein
Studium in 6 Semestern und promovierte.

Für die Wirksamkeit ERTELS sollten weitere Umstände günstig sein:
FICKER war Sekretär und Mitglied der Berliner Akademie. Hinzu
kommt, dass sowohl ALBERT DEFANT (1884–1974) sowie JULIUS
BARTELS (1899–1964) der Akademie angehörten. Alle waren bestens
bekannt mit MAX VON LAUE (1879–1960), ALBERT EINSTEIN (1879–
1955), MAX PLANCK (1958–1947), ERWIN SCHRÖDINGER (1887–
1961) u. a., denen auch ERTEL vorgestellt wurde. Diese Gelehrten
wurden rasch auf ERTEL aufmerksam und förderten ihn nach Kräften.
So wundert es nicht, dass ERTEL Themen auch aus dem Umfeld
EINSTEINS und SCHRÖDINGERS aufgriff. Dies alles waren aber Nebenschritte, denn ERTEL kehrte zu FICKER zurück und widmete sich
ganz der theoretischen Meteorologie.

Aus dem eingangs erwähnten Briefwechsel BJERKNES-FICKER war
die Ähnlichkeit, aber auch der Gegensatz beider Persönlichkeiten
deutlich geworden. In die nachfolgende Zeit fiel die Diskussion um
die Steuerung des Wettergeschehens durch die Stratosphäre, die
Frage der Zusammensetzung der Depressionen sowie die weitergehende Ausarbeitung des BJERKNESschen Konzeptes. Anlässlich der
17. Versammlung der Deutschen Meteorologischen Gesellschaft in
Wien 1930 hatte sich ERTEL mit diesem Komplex auseinandergesetzt.
Kurz danach räumte BJERKNES in seiner „Physikalischen Hydrodynamik" ausdrücklich die Priorität FICKERS bei der Entdeckung
der großen Diskontinuitäten für die Wetterentwicklung ein. BJERKNES
betonte, dass es die norwegische Schule versäumt hatte, den
FICKERschen Ergebnissen rechtzeitig zu folgen. FICKER selbst hatte
1935 nochmals die Frage des Einflusses der Stratosphäre auf die
Wetterentwicklung aufgegriffen. In der weiteren Diskussion hat ERTEL
insofern eine Rolle gespielt, als er verschiedene theoretische Fragen,
wie z. B. zur zonalen Zirkulation, den atmosphärischen Druckschwankungen des Windfeldes an der Tropopause, der Zyklonenbewegung
usw., aufgegriffen hat. In zwei Monographien hat ERTEL das
theoretische Gebäude der Meteorologie umrissen: „Methoden und

Probleme der dynamischen Meteorologie" (1938) und „Die theoretischen Grundlagen der Meteorologie" (1939). In beiden Bänden würdigt ERTEL den norwegischen Beitrag, aber auch FICKERS Beiträge werden hinreichend dargestellt.

Es wurde bereits erwähnt, dass das Meteorologische Institut in Berlin seit 1936 eine eigene Reihe „Veröffentlichungen ..." herausgab. Als Herausgeber zeichneten FICKER und ERTEL. Das erste Heft dieser Reihe ist die ERTELsche Abhandlung „Advektiv-dynamische Theorie der Luftdruckschwankungen und ihrer Periodizitäten" (1939). Die weiteren Hefte beinhalten u. a. Beiträge von K. BROCKS, H. FICKER, J. JAW, S. LI, O. SCHNEIDER, K. STUMPFF sowie I. WEISS. Von diesen Mitarbeitern hatte BROCKS später in Hamburg eine leitende Funktion inne, SCHNEIDER war Vorsitzender des argentinischen IGY- und Antarktis-Programms und JAW hatte eine führende Rolle in China inne. Wiederholt wurde ERTEL später auch von der Chinesischen Akademie der Wissenschaften eingeladen, da ihm in China eine ganz besondere Wertschätzung und Verehrung galt.

Im Jahre 1937 übernahm FICKER das Direktorat der Zentralanstalt für Meteorologie und Geodynamik in Wien sowie die Professur für Physik der Erde an der Wiener Universität. In dieser Zeit riss aber die freundschaftliche Verbindung zwischen FICKER und ERTEL nicht ab.

Abb. 1. Heinrich Ficker   Abb. 2. Hans Ertel
(Bild im Privatbesitz)   (Bild im Privatbesitz)

ERTEL erhielt eine Planstelle an der Zentralanstalt für Meteorologie in Wien, trat sie aber niemals an, da er sofort für die Meteorologenausbildung in Berlin abgeordnet wurde. Zwischen FICKER und ERTEL bestanden vor allem briefliche Kontakte, jedoch hat FICKER nachdrücklich geholfen, dass ERTEL 1943 Ordinarius an der Universität Innsbruck wurde. Diese Stelle musste er nach 1945 räumen; sie wurde von ALBERT DEFANT eingenommen.

Kurz danach nahm ERTEL am internationalen Projekt der isentropen Analyse am MIT in Cambridge (USA) teil, wobei sich seine Beziehungen zu CARL-GUSTAF ROSSBY wesentlich vertiefen sollten. Die Freundschaft beider Gelehrter hielt ein Leben lang an, und auch später trafen sich beide noch, so u. a. bei der Jubiläumsfeier der Zentralanstalt für Meteorologie in Wien 1950. Später versuchte ROSSBY immer wieder, ERTEL in internationale Forschungsvorhaben der Meteorologie einzubinden. Dass dies nicht gelang, lag sicher daran, dass ERTEL wegen seiner Tätigkeit als Vizepräsident der Deutschen Akademie der Wissenschaften u. a. einfach keine Zeit mehr übrig hatte für weitere Verpflichtungen.

Die weitere Zusammenarbeit zwischen ERTEL und FICKER war seit dessen Weggang im Brieflichen zu suchen. Von FICKER wird berichtet, dass er gerne alle schwierigen Probleme sammelte und diese dann an ERTEL schickte. Dann dauerte es, bis er alle erhofften Lösungen erhielt. Zum 60. Geburtstag von H. FICKER veröffentlichte ERTEL 1941 in den „Naturwissenschaften" einen sehr lesenswerten Beitrag (ERTEL [8]). Zwar bemühte sich FICKER darum, ERTEL nach Wien zu bekommen, doch dieser wollte nicht. ERTEL hatte schon damals alle anderen akademischen Möglichkeiten außerhalb Berlins abgelehnt. Lediglich 1943 nahm ERTEL einen Ruf nach Innsbruck an, den er dann 1945 abgeben musste (s. o.). Auch später hat er Rufe z. B. nach Graz, München und Wien nicht angenommen. Besonders um die Berufung nach Graz sowie Wien hatte sich FICKER bemüht. Er hätte es gerne gesehen, dass z. B. ERTEL sein Nachfolger in Wien werde. Übrigens scheiterte die Berufung nach Graz auf postalischem Wege: Der Brief wurde ERTEL viel zu spät zugestellt und die Stelle anderweitig besetzt. Wenngleich FICKER ERTEL gerne an der Universität Wien gesehen hätte, so wollte dieser Berlin nicht verlassen.

Dessen ungeachtet waren die Beziehungen ERTEL-FICKER in all den Jahren sehr herzlich. Ein Wiedersehen gab es auch anlässlich der 250-Jahr-Feier der Deutschen Akademie der Wissenschaften, an der eine Delegation der Österreichischen Akademie der Wissenschaften teilnahm, so u. a. der Generalsekretär KEIL und FICKER. FICKER vertrat dabei zugleich auch die Wiener Universität. Stets war FICKER auch daran interessiert, z. B. über ERTELS Beziehung zu ROSSBY zu erfahren. So fragte er ERTEL u. a. in einem Brief vom 29. Januar 1950:

„Es würde mich sehr interessieren, von Dir zu erfahren, wie der Besuch ROSSBYS außerhalb des Bräustübls ausgefallen ist. Ich kann mich in die neue Meteorologie nicht mehr hineinfinden, wohl auch deshalb, weil noch keine gute Darstellung in deutscher Sprache vorhanden ist. Es berührt mich fast komisch, dass trotz der Ausbildung von REUTER (REUTER war später Professor für theoretische Meteorologie in Wien) in Stockholm und dem Kurs Dr. KLETTERS (Mitarbeiter von FICKER) in Kissingen und ihres Bemühens, in der Synoptik (praktischen Wettervorhersage) nach SCHERHAG und ROSSBY zu arbeiten, in der Praxis sich doch meine alten Bauernregeln am besten bewähren."

In diesem Zusammenhang mag auch an ein Schreiben von DEFANT an ERTEL vom 17. Juli 1954 erinnert werden, als sich dieser für die Glückwünsche der Deutschen Akademie der Wissenschaften zu seinem 70. Geburtstag bedankt, die ERTEL verfasst hatte. Er schreibt u. a.:

„Ihre Zeilen haben mir die schönen Zeiten meiner erfolgreichsten Jahre in Berlin lebhaft in Erinnerung gebracht und des Zusammenseins mit Ihnen."

Mit der Wahl zum korrespondierenden Mitglied der Österreichischen Akademie der Wissenschaften hatten sich ERTELS Beziehungen zu Österreich weiter vertieft. In dem Wahlvorschlag (1956) lobten FICKER u. a. ERTEL als den besten Theoretiker der Meteorologie im deutschen Sprachraum. ERTEL selbst war über die Wahl sehr erfreut und hatte später bei einer Tagung 1965 nochmals Gelegenheit, Wien wiederzusehen und Freunde der Akademie, u. a. A. DEFANT und F. STEINHAUSER, zu treffen. FICKERS Tod (1957) hat ERTEL sehr betroffen gemacht. In einem ersten Nachruf (ERTEL 1958) schrieb er:

„Mit tiefem Schmerz empfinden wir Meteorologen den Verlust dieses Mannes, in dessen ganzem Wesen die Eigenschaften eines klarblickenden Forschers mit der warmen menschlichen Herzensgüte eines edlen Charakters harmonisch zusammenklangen" (Zeitschrift für Meteorologie 1958).

An die Österreichische Akademie der Wissenschaften schickte die Deutsche Akademie der Wissenschaften folgenden Brief:

„Die Deutsche Akademie der Wissenschaften zu Berlin bringt in tiefer Trauer der Österreichischen Akademie der Wissenschaften zum Hinscheiden ihres Vizepräsidenten und Präsidenten der mathematisch-naturwissenschaftlichen Klasse Prof. em. Dr. Dr. h.c. HEINRICH FICKER ihr aufrichtiges Beileid zum Ausdruck. Mit dem Dahingeschiedenen hat die meteorologische Wissenschaft einen Forscher von höchstem Rang und einen Altmeister ihres Faches verloren, der mit begeistertem Streben nach Erkenntnis der Natur schlichte Vornehmheit seines Charakters, Lauterkeit seiner Gesinnung und tiefinnige Herzensgüte harmonisch verband. Die Deutsche Akademie der Wissenschaften zu Berlin wird diesem hochverdienten Gelehrten und guten Menschen dauernd ein ehrendes Gedenken bewahren. In aufrichtiger Anteilnahme, HANS ERTEL."

Das waren Worte, die auch für ERTEL selbst galten. Im Jahrbuch der Deutschen Akademie der Wissenschaften für 1962 hat ERTEL dann einen ausführlicheren Nachruf geschrieben (ERTEL [10]).

Fast dreißig Jahre verband FICKER und ERTEL eine besondere, im Laufe der Zeit tief gewachsene Freundschaft, die ihresgleichen sucht. FICKER hat selbstlos den wissenschaftlichen Werdegang ERTELS gefördert. Er hat ihn bekannt gemacht und eingeführt in die wissenschaftliche Welt EINSTEINS, PLANCKS, LAUES, SCHRÖDINGERS u. a., die ihrerseits ERTEL mit Herzlichkeit aufnahmen und umfassend förderten. So konnte sich ERTEL in beispielloser Weise unter der Obhut und Güte weltbekannter Wissenschaftler zu einem hervorragenden Gelehrten entwickeln.

Es mögen noch ein paar Worte angefügt werden von Prof. HEINZ REUTER (Wien), der ein Schüler FICKERS war. Er schrieb mir in einem Brief:

„Nun noch ein paar Worte über FICKER. Er hat mir das Verständnis für die praktische Meteorologie beigebracht. Ich habe ihn immer schon deshalb bewundert, weil er ohne besondere theoretische Kenntnisse die Vorgänge in der Natur beschreiben und klären konnte, was bei der Komplexität der atmosphärischen Prozesse einmalig erscheint. Auch MAX PLANCK soll (so hat mir ERTEL einmal erzählt) gesagt haben, dass er keinen anderen Naturwissenschaftler kennengelernt habe, der fähig gewesen wäre, ohne Hilfe der Theorie Prozessabläufe in der Natur zu beschreiben, und zwar nach einem strengen Kausalprinzip. Wenn ich zurückblicke in die Zeit meiner ersten wissenschaftlichen meteorologischen Arbeiten und die ganze (rasante) Entwicklung unserer Wissenschaft bis heute an meinem geistigen Auge vorbeiziehen lasse, dann glaube ich, behaupten zu können, dass ERTEL die theoretischen Grundlagen für die bis heute so weit entwickelte Theorie geliefert hat und FICKER der wahrscheinlich letzte empirische große Synoptiker war." (REUTER [13].)

3. Ausblick

Es ist wiederholt, auch in der Meteorologie, von wissenschaftlichen Schulen die Rede gewesen. Man spricht von der Chicagoer Schule, von der norwegischen Meteorologenschule, so dass gefragt werden kann, wie es im deutschen Sprachraum ist. Sicher hat FICKER, aber auch ERTEL, keine wissenschaftliche Schule gebildet. Beide waren sich einig darin, die individuellen Möglichkeiten der ihnen anvertrauten Menschen zu fördern, so gut es ging. Gerade in der Bandbreite der Begabungen sahen sie die Chance für den wissenschaftlichen Werdegang, und so wundert es nicht, dass aus ihrem Zuhörerkreis viele Themen behandelt wurden, die die gesamte Geophysik und Meteorologie abdecken. Gerade darin lag die Ausrichtung dieser beiden akademischen Lehrer. Die Begabungsvielfalt war es, die gefördert wurde.

Später hat sich ERTEL, nach Gründung des Instituts für Physikalische Hydrographie, besonders den praktischen Problemen zugewandt. Das Institut, eine Zeit lang international führend in verschiedenen Teilgebieten der Hydrodynamik und physikalischen Hydrographie, gab außerdem ab 1953 eine eigene Schriftenreihe heraus: die Acta Hydrophysica. Sie sollte die Theoretiker und Praktiker zusammenführen, wie dies schon LEIBNIZ gefordert hatte. Im Laufe der Jahre – ERTEL gab die Zeitschrift bis 1971 heraus – erschienen auch viele Arbeiten, die sowohl den theoretischen Aspekt als auch den praktischen Nutzen für die Volkswirtschaft berücksichtigten. Überdies wurde die Zeitschrift als Tauschobjekt mit anderen Institutionen in aller Welt genutzt, wodurch ERTELS Institut sehr viele ausländische Bücher und Zeitschriften erhielt, etwas, was damals durchaus sehr wichtig war. Interessant ist auch, dass, fast an FICKERS praktische Richtung erinnernd, auch ERTEL die Forschungsthematik des Instituts praxisorientiert sah. Lange bevor überhaupt in der Geophysik von ökologischen Fragestellungen gesprochen wurde, hatte das Institut für Physikalische Hydrographie in Berlin international bedeutende Probleme der Geo-Ökologie sowie des Umweltschutzes in den Grundzügen bearbeitet. Hierzu zählt auch die vom Institut herausgegebene „Quellensammlung zur Hydrographie und Meteorologie", die CURT WEIKINN, der von ERTEL ans Institut geholt worden war, erfolgreich zusammenstellen konnte. Das Werk hat höchste internationale Anerkennung gefunden, so z. B. in Rezensionen von H. E. LANDSBERG (USA), H. LAMB (England) und A. RETHLY (Ungarn) u. a.

Für die Entwicklung der Meteorologie war die Begegnung von FICKER mit ERTEL ein Glücksfall. Dadurch, dass FICKER ERTEL sozusagen „entdeckte", dessen Werdegang positiv begleitete und ERTEL sich somit umfassend entwickeln konnte, wuchs der Erkenntniszuwachs in den meteorologischen und geophysikalischen Disziplinen. FICKERS und ERTELS Beziehung ist ein Beispiel der harmonischen Entwicklung eines gütigen Lehrers für seinen

Schüler, wobei sich dieser ganz im Geiste FICKERS ebenso gegenüber allen verhielt, die sich auf den Weg der Wissenschaft
machten.

Ungedruckte Quellen

Die Briefe FICKER/BJERKNES befanden sich vor Jahren in der Handschriftenabteilung
der UB Oslo. Die Briefe ERTEL/FICKER sind im Privatbesitz. ERTELS Brief sowie der
Wahlvorschlag an die ÖAW befinden sich in deren Archiv (Personalakt HEINRICH
VON FICKER).

Danksagung

Der Universitätsbibliothek Oslo sowie dem Archiv der Österreichischen Akademie der
Wissenschaften danke ich für freundliche Hilfe. Herrn Professor HELMUT MORITZ
(Graz) danke ich, dass er mir nachdrücklich bei Verbesserungen half, für die er sich
wieder Rat bei Professor HEINZ KAUTZLEBEN (Berlin) holte.

Bemerkung

Eine Vorveröffentlichung erschien im Internet unter http://verplant.org/history-
geophysics/Ficker.htm. Die Leibniz-Sozietät der Wissenschaften in Berlin hat in
gleicher Weise des hervorragenden Meteorologen HEINRICH FICKER gedacht. Eine
andere Fassung dieses Beitrages wurde der Leibniz-Sozietät vorgelegt.

Literatur

[1] BJERKNES, V., BJERKNES, J., SOLBERG, H., BERGERON, T. (1933) Physikalische
     Hydrodynamik mit Anwendung auf die dynamische Meteorologie. Springer,
     Berlin
[2] ERTEL, H. (1932) Theorie der durch Variationen des magnetischen Potentials
     induzierten Erdströme bei ungleichförmiger Leitfähigkeit der Erdrinde (Inau-
     gurationsdiss. 1932). Archiv f. Erdmagnetismus Heft 8
[3] ERTEL, H. (1931) Der Einfluß der Stratosphäre auf die Dynamik des Wetters.
     Meteorol. Z. **48**: 461
[4] ERTEL, H. (1936) Singuläre Advektion und ihre Darstellung durch C. G. Rossbys
     Advektionsfunktion. Veröff. Meteorol. Inst. U. Berlin I, Heft 6
[5] ERTEL, H. (1938) Methoden und Probleme der dynamischen Meteorologie.
     Springer, Berlin
[6] ERTEL, H. (1939) Die theoretischen Grundlagen der dynamischen Meteorologie.
     Meteorol. Taschenbuch, V. Ausg., herausg. von F. LINKE. Bahrt, Leipzig
[7] ERTEL, H. (1940) Elemente der Operatorenrechnung mit geophysikalischen
     Anwendungen. Springer, Berlin
[8] ERTEL, H. (1941) Heinrich von Ficker zu seinem 60. Geburtstag am 22.
     November 1941. Naturwiss. **29**: 697
[9] ERTEL, H. (1957) Schreiben an die Österreichische Akademie der
     Wissenschaften (Archiv Österr. Akad. Wiss., Personalakt Heinrich von Ficker,
     Mappe 1)

[10] ERTEL, H. (1963) Heinrich Ficker. Jahrbuch der Deutschen Akad. Wiss. zu Berlin, 1962. Akademie-Verlag, Berlin

[11] FICKER, H. (1935) Der Einfluß der Stratosphäre auf die Wetterentwicklung. Naturwiss. **23**: 552

[12] PLANCK, M. (1948) Max Planck in seinen Akademie-Ansprachen. Erinnerungsschrift der Deutschen Akademie der Wissenschaften zu Berlin. Akademie-Verlag, Berlin

[13] REUTER, H. (1987) Brief an den Verfasser vom 12. 12. 1987

[14] SCHRÖDER, W. (1971) Nachruf auf Hans Ertel. Wetter und Leben, Heft 4

Anschrift des Verfassers: Dr. Wilfried Schröder, Geophysikalische Station, Hechelstraße 8, 28777 Bremen, Deutschland.

Österreichische Akademie der Wissenschaften
Mathematisch-naturwissenschaftliche Klasse

Sitzungsberichte

Abteilung II

Mathematische, Physikalische
und Technische Wissenschaften

216. Band
Jahrgang 2007

Wien 2008

Verlag der Österreichischen Akademie der Wissenschaften

Inhalt

Sitzungsberichte Abt. II

Sitzungsber. Abt. II (2007) 216: 3–13

Sitzungsberichte

Mathematisch-naturwissenschaftliche Klasse Abt. II
Mathematische, Physikalische und Technische Wissenschaften

On Some Functional Equations Involving Involutions

By

Andrzej Mach and Zenon Moszner

(Vorgelegt in der Sitzung der math.-nat. Klasse am 22. März 2007
durch das w. M. Ludwig Reich)

Abstract

In this note we present some theorems characterizing solutions of the equation
$f(x) = f(\varphi(x)) + g(x)$, where φ is a given involution, and particulary differentiable
solutions of the equation $f(x) = f(1 - x) + 2x - 1$. The stability of this equation and
nonexistence of the extremal points of the set of solutions are proved.

Mathematics Subject Classification (2000): 39B72.
Key words: Functional equations, involution, stability, extremal points.

1. Introduction

WŁODZIMIERZ WYSOCKI, in connection with the problems of copulas,
considers the fixed points of the application $f \mapsto f(1 - x) + 2x - 1$ in
the class $\mathcal{F}$ of functions $f \colon [0, 1] \to \mathbf{R}$, such that $f(0) = f'_+(0) = 0$,
$f(1) = 1, f'_-(1) = 2, f''(x) > 0$, for $x \in \,]0, 1[$ (thus the functions f are
the bijections on $[0, 1]$). This consideration gives the functional
equation

$$f(x) = f(1 - x) + 2x - 1 \tag{1}$$

and the generalization of this equation, namely

$$f(x) = f(\varphi(x)) + g(x), \tag{2}$$

where $\varphi: E \to E$ is a given involution, this means

$$\varphi(\varphi(x)) = x, \tag{3}$$

$f, g: E \to G$ are unknown functions, E is an arbitrary set and $(G, +)$ is an arbitrary group.

In this note we present some theorems characterizing solutions of Eqs. (1) and (2) and particularly, the differentiable solutions of Eq. (1).

It is proved the stability of Eq. (2) and of the equation

$$g(\varphi(x)) + g(x) = 0. \tag{4}$$

Moreover, we obtain the nonexistence of the extremal points of the set of solutions of (2).

2. General Case

2.1. General Solution

Let $\varphi: E \to E$ be an involution.

Definition. We define the equivalence relation $\rho \subset E \times E$ by the following formula

$$x \rho y \Leftrightarrow y = \varphi(x) \quad \text{or} \quad y = x.$$

By E_1 we denote an arbitrary selection of the set E/ρ and let $E_2 := E \backslash E_1$ and $E_3 := E \backslash (E_2 \cup \varphi(E_2)) = \{x \in E : \varphi(x) = x\}$.

Theorem 1. *The functions $f, g: E \to G$ are the solutions of Eq. (2) if and only if*

$$f(x) = \begin{cases} a(x) & \text{for} \quad x \in E_1, \\ a(\varphi(x)) + b(x) & \text{for} \quad x \in E_2, \end{cases} \tag{5}$$

$$g(x) = \begin{cases} b(x) & \text{for} \quad x \in E_2, \\ -b(\varphi(x)) & \text{for} \quad x \in \varphi(E_2), \\ 0 & \text{for} \quad x \in E_3, \end{cases} \tag{6}$$

where $a: E_1 \to G$, $b: E_2 \to G$ are the arbitrary functions.

Moreover, the function $g(x)$ given by formula (6) is a solution of Eq. (4).

Easy calculations show that functions defined by (5) and (6) satisfy (2) and the function (6) fulfils (4).

Proof of "only if". We suppose that the functions f, g satisfy Eq. (2). We define $a(x) := f(x)$ for $x \in E_1$ and $b(x) := g(x)$ for $x \in E_2$. If $x \in E_2$

then we have by Eq. (2)

$$f(x) = f(\varphi(x)) + g(x) = a(\varphi(x)) + b(x).$$

This means that the function f has the form (5). If $x \in \varphi(E_2)$ then we have $f(x) = a(x), f(\varphi(x)) = a(x) + b(\varphi(x))$ and by Eq. (2)

$$a(x) = a(x) + b(\varphi(x)) + g(x),$$

so $g(x) = -b(\varphi(x))$. If $x \in E_3$ then we have $f(x) = a(x), f(\varphi(x)) = a(x)$ and by Eq. (2)

$$a(x) = a(x) + g(x),$$

so $g(x) = 0.$ $\qquad \square$

It turns out that the general solution of (4) is given by formula (6) only in the case where the group G has no element of the order 2. More precisely, we have the following result, the proof of which is similar to that of Theorem 1.

Theorem 2. *The function g given by the formula*

$$g(x) = \begin{cases} b(x) & for \quad x \in E_2, \\ -b(\varphi(x)) & for \quad x \in \varphi(E_2), \\ c(x) & for \quad x \in E_3, \end{cases}$$

where $b: E_2 \to G$, $c: E_3 \to G_1 := \{\alpha \in G: 2\alpha = 0\}$ are the arbitrary functions, is the general solution of Eq. (4).

The proof of the below presented theorem is evident.

Theorem 3. *The general solution of Eq. (3) has the form*

$$\varphi(x) = \begin{cases} \psi(x) & for \quad x \in Z_2, \\ \psi^{-1}(x) & for \quad x \in \psi(Z_2), \\ x & for \quad x \in E \setminus (Z_2 \cup \psi(Z_2)), \end{cases} \qquad (7)$$

where $Z_2 \subset E$ is such that $\operatorname{card} Z_2 \leq \operatorname{card}(E \setminus Z_2)$ and $\psi: Z_2 \to E \setminus Z_2$ is an arbitrary injection. Thus the general solution of the system of Eqs. (2) and (3) is of the form (7),

$$f(x) = \begin{cases} a(x) & for \quad x \in Z_1 := E \setminus Z_2, \\ a(\psi(x)) + b(x) & for \quad x \in Z_2, \end{cases}$$

$$g(x) = \begin{cases} b(x) & for \quad x \in Z_2, \\ -b(\psi^{-1}(x)) & for \quad x \in \psi(Z_2), \\ 0 & for \quad x \in E \setminus (Z_2 \cup \psi(Z_2)), \end{cases}$$

where Z_2 and ψ are as above and $a: Z_1 \to G$, $b: Z_2 \to G$ are the arbitrary functions.

In the next result (2) is considered as an equation with an unknown function f and a given function g.

Theorem 4. (1^0) *Equation* (2) *has a solution if and only if g satisfies* (4).

Assume that $g, g_1: E \to G$ satisfy (4).

(2^0) *If $f_1: E \to G$ is a solution of* (2), *and $\Phi: G \to G$ is an even function, then the function $f: E \to G$ given by*

$$f = \Phi \circ g_1 + f_1 \tag{8}$$

satisfies (2).

(3^0) *If g_1 is invertible then the general solution $f: E \to G$ of* (2) *can be obtained in the following way. Fix a solution $f_1: E \to G$ of* (2), *choose an even function $\Phi: g_1(E) \to G$ and put* (8).

(4^0) *The general solution of* (2) *is a sum of general solutions of the equation*

$$m(\varphi(x)) = m(x) \tag{9}$$

and of the particular solution of (2). *The general solution of Eq.* (9), *for arbitrary sets E and G, is of the form*

$$m(x) = \begin{cases} d(x) & \text{for} \quad x \in E_1, \\ d(\varphi(x)) & \text{for} \quad x \in E_2, \end{cases}$$

where $d: E_1 \to G$ is an arbitrary function.

Proof. The proof in the cases (1^0), (2^0) and (4^0) is obvious. In case (3^0), the function $\Phi(u) = f[g_1^{-1}(u)] - f_1[g_1^{-1}(u)]$, for some solutions f and f_1 of (2), is even. Indeed, since $g_1(\varphi(x)) = -g_1(x)$, then $\varphi[g_1^{-1}(u)] = g_1^{-1}(-u)$, the set $g_1(E)$ is symmetric in relation to 0 (if $a = g_1(x)$ then $-a = g_1(\varphi(x))$) and

$$\Phi(-u) = f[g_1^{-1}(-u)] - f_1[g_1^{-1}(-u)] = f\{\varphi[g_1^{-1}(u)]\} - f_1\{\varphi[g_1^{-1}(u)]\}$$
$$= f[g_1^{-1}(u)] - g(g_1^{-1}(u)) - \{f_1[g_1^{-1}(u)] - g(g_1^{-1}(u))\}$$
$$= \Phi(u) \qquad \text{for} \quad u \in g_1(E).$$

$\square$

Remark 1. Every even function $\Psi: g_1(E) \to G$ is obviously extended to the even function on G since $g_1(E)$ is symmetric in relation to 0, there it is possible to use in (3^0) for Φ the even function on G (not only on $g_1(E)$).

Remark 2. The invertible solution of (4) cannot exist, e.g., for $E = \mathbf{R}$, $(G, +) = (\mathbf{R}, +)$, $\varphi(x) = x$. This solution exists if and only if $\operatorname{card} E_2 \leq \operatorname{card}(G \backslash G_1)/R$ and $\operatorname{card} E_3 \leq \operatorname{card} G_1$, where for $\alpha, \beta \in G \backslash G_1$: $\alpha R \beta \Leftrightarrow (\beta = -\alpha$ or $\beta = \alpha)$ (see Theorem 2). The invertible solutions of (4) are determined exactly to an invertible, odd function. For two invertible solutions g_1 and g_2 of (4) we have

$$g_2[g_1^{-1}(-u)] = g_2[\varphi(g_1^{-1}(u))] = -g_2[g_1^{-1}(u)] \qquad \text{for} \quad u \in g_1(E),$$

thus $g_2 = \Psi(g_1)$, where $\Psi: g_1(E) \to G$ is invertible and odd. Inversely, if g_1 is an invertible solution of (4), then $g_2 = \Psi(g_1)$ is the same solution, where $\Psi: g_1(E) \to G$ is an arbitrary, invertible and odd function.

Example for (3^0) in Theorem 4 $(g(x) = \operatorname{sgn}(2x - 1), g_1(x) = 2x - 1)$. For the equation

$$f(x) = f(1 - x) + \operatorname{sgn}(2x - 1),$$

where $f: \mathbf{R} \to \mathbf{R}$ and $\operatorname{sgn} 0 := 0$, $f_1(x) = 1$ for $x \geq \frac{1}{2}$ and $f_1(x) = 0$ for $x < \frac{1}{2}$ is a particular solution, $\Phi(2x - 1) + f_1(x)$ for the arbitrary even function $\Phi: \mathbf{R} \to \mathbf{R}$ is a general solution and $\Phi[\operatorname{sgn}(2x - 1)] + f_1(x)$ is a solution but not the general solution. Indeed, this function is for $x > \frac{1}{2}$ equal to $\Phi(1) + 1$ and the function $(2x - 1)^2 + f_1(x)$ is a solution of our equation and it is not constant for $x > \frac{1}{2}$. Therefore, the assumption that g_1 is invertible is essential in (3^0) of Theorem 4.

2.2. Stability

We will prove that the system of Eqs. (2) and (4), Eq. (2) and Eq. (4), are stables for some metrics.

Theorem 5. (1^0) *Let us suppose that ρ is the left translation-invariant metric in the group $(G, +)$, this means $\rho(u, v) = \rho(t + u, t + v)$, for $u, v, t \in G$. For every $\varepsilon > 0$ and for every functions $a: E \to G$, $b: E \to G$ such that*

$$\forall x \in E: \qquad \rho(a(\varphi(x)) + b(x), a(x)) \leq \varepsilon, \qquad (10)$$

$$\forall x \in E: \qquad \rho(b(\varphi(x)) + b(x), 0) \leq \varepsilon, \qquad (11)$$

there exists a solution $f: E \to G$, $g: E \to G$ of the system of Eqs. (2) and (4) such that

$$\forall x \in E: \qquad \rho(f(x), a(x)) \leq \varepsilon \qquad (12)$$

and

$$\forall x \in E: \qquad \rho(g(x), b(x)) \leq \varepsilon. \qquad (13)$$

(2^0) *If the metric ρ is the left and right translation invariant and only the inequality (10) is true, then (12) is satisfied and we have (13) with 2ε in place of ε.*

(3^0) *If the metric ρ is the left translation invariant, $2\rho(u,0) \leq \rho(2u,0)$, for $u \in G$ and only (11) is true, then (13) is satisfied.*

Proof. Functions f, g given by (5), (6) satisfy (2) and (4), by Theorem 1. Hence:

If $x \in E_2$, then $\rho[f(x), a(x)] = \rho[a(\varphi(x)) + b(x), a(x)]$ and $\rho[g(x), b(x)] = 0$.

If $x \in E_3$, then $\rho[f(x), a(x)] = 0$, $\rho[g(x), b(x)] = \rho[0, b(x)] = \rho[a(\varphi(x)), a(\varphi(x)) + b(x)] = \rho[a(x), a(\varphi(x)) + b(x)]$ and $\rho[0, b(x)] \leq \frac{1}{2}\rho[0, 2b(x)]$ in case (3^0).

If $x \in \varphi(E_2)$, then $\rho[f(x), a(x)] = 0$ and $\rho[g(x), b(x)] = \rho[-b(\varphi(x)), b(x)] = \rho[0, b(\varphi(x)) + b(x)] = \rho[a(x), a(x) + b(\varphi(x)) + b(x)] \leq \rho[a(x), a(\varphi(x)) + b(x)] + \rho[a(\varphi(x)), a(x) + b(\varphi(x))]$. $\square$

Remark 3. The equation

$$g(x) + g(x) = 2g(x) = 0 \quad (\varphi(x) = x) \tag{14}$$

is not stable for some metric.

Let be

$$\alpha(x) = \begin{cases} x & \text{for} \quad x \in Z_1 := \{2/(2n+1): n = 1, 2, \ldots\}, \\ 1 + x & \text{for} \quad x \in Z_2 := \{1/(2n+1): n = 1, 2, \ldots\}, \\ \beta(x) & \text{for} \quad x \in Z_3 := \mathbf{R}\backslash(Z_1 \cup Z_2), \end{cases}$$

where $\beta: Z_3 \to \mathbf{R}\backslash[Z_1 \cup (Z_2 + 1)]$ is a bijection, such that $\beta(0) = 0$. Let $\rho(a, b) := |\alpha(a) - \alpha(b)|$ be the metric in $\mathbf{R}$. Let $(G, +) = (\mathbf{R}, +)$ and $E = \{0\}$. Eq. (14) is not stable with respect to the metric above. Indeed, for $\varepsilon = 1$ and $\delta > 0$, if $2/(2n+1) \leq \delta$ and $k(0) = 1/(2n+1)$, then

$$\rho[2k(0), 0] = |\alpha(2/(2n+1)) - \alpha(0)| = 2/(2n+1) \leq \delta$$

and

$$\rho[k(0), 0] = |\alpha(1/(2n+1)) - \alpha(0)| = 1 + 1/(2n+1) > 1.$$

Attention. Here the equation $2g(x) = 0$ is not stable and the equivalent equation $g(x) = 0$ is evidently stable.

Remark 4. The system of Eqs. (2), (3), (4) is not stable for some metrics in E and G, since *the equation of involution is not stable for some translation-invariant metric ρ^* in E,* i.e., the condition below (the analogon of the Ulam-Hyers stability of the equation of homo-morphism in [1]) is not satisfied:

For every $\varepsilon > 0$ there exists a $\delta > 0$ such that for every $\psi: E \to E$ if $\rho^*(\psi(\psi(x)), x) \leq \delta$ for $x \in E$, then there exists an involution φ such that $\rho^*(\varphi(x), \psi(x)) \leq \varepsilon$ for $x \in E$.

Indeed, if $E = \mathbf{R}$ with natural metric, then for arbitrary $\delta > 0$ and n fixed such that $1/n \leq \delta$ and $n > 3$, the function $\psi: E \to E$ such that $\psi(0) = \psi(1/n) = 1$, $\psi(1) = 1/n$, $\psi(\alpha) = \alpha$, for $\alpha \in E \setminus \{0, 1, 1/n\} =: E^*$, satisfies $|\psi(\psi(x)) - x| \leq \delta$, for $x \in E$. We suppose that there exists an involution φ such that $|\psi(x) - \varphi(x)| \leq \frac{1}{3}$, for $x \in E$. Then $|\psi(0) - \varphi(0)| = |1 - \varphi(0)| \leq \frac{1}{3}$, thus $\frac{2}{3} \leq a := \varphi(0) \leq \frac{4}{3}$. We have $|\psi(a) - \varphi(a)| = |\psi(a) - 0| = |\psi(a)| \leq \frac{1}{3}$, thus $a = 0$ or $a = 1/n$ is impossible. If $a \in E^*$, then $|a| \leq \frac{1}{3}$, thus a contra-diction. Therefore, $\varphi(0) = a = 1$. We have $|\psi(1/n) - \varphi(1/n)| = |1 - \varphi(1/n)| \leq \frac{1}{3}$, thus $\frac{2}{3} \leq b := \varphi(1/n) \leq \frac{4}{3}$ and $|\psi(b) - (1/n)| \leq \frac{1}{3}$. For $b \in E^*$: $|b - (1/n)| \leq \frac{1}{3}$, thus $b \leq (1/n) + \frac{1}{3} < \frac{2}{3}$, we obtain a con-tradiction. For $b = 0$ or $b = 1/n$ the contradiction is evident. Therefore, $\varphi(1/n) = b = 1$. A contradiction, since φ, as involution, is injective.

If $(G, +)$ is the group with arbitrary metric ρ, then the triple $(h(x), k(x), \psi(x)) = (0, 0, \psi(x))$ satisfies the inequalities $\rho[h(\psi(x)) + k(x), h(x)] \leq \delta$, $\rho[k(\psi(x)) + k(x), 0] \leq \delta$, $|\psi(\psi(x)) - x| \leq \delta$, for $x \in E$, and the solution (f, g, φ) of the system: (2), (4), (3) such that $|\psi(x) - \varphi(x)| \leq \frac{1}{3}$, for $x \in E$, does not exist.

If the metric is $0 - 1$ ($\rho(x, x) = 0$ and $\rho(x, y) = 1$ for $x \neq y$) then every functional equation (or system) is stable (it is sufficient to put arbitrary positive $\delta < 1$ for all $\varepsilon > 0$).

3. Particular Case. Differentiable Solutions

By Theorem 4 we have the following corollaries.

Corollary 1. *The function $f: \mathbf{R} \to \mathbf{R}$ given by the formula*

$$f(x) = x^2 + \Phi(2x - 1), \tag{15}$$

where $\Phi: \mathbf{R} \to \mathbf{R}$ is an arbitrary even function, is the general solution of the functional equation (1) without any regularity condition.

The differentiable function $f: \mathbf{R} \to \mathbf{R}$ is the solution of the functional equation (1), for which $f'(0) = 0$ and $f'(1) = 2$, if and only if there exists a differentiable and even function $\Phi: \mathbf{R} \to \mathbf{R}$, such that $\Phi'(-1) = \Phi'(1) = 0$ and (15) is satisfied.

Corollary 2. *The function $f: [0, 1] \to \mathbf{R}$ belonging to the class $\mathcal{F}$ is the solution of the functional equation (1), if and only if there exists a* *twice differentiable and even function Φ defined on interval $[-1, +1]$, such that*

$$-\tfrac{1}{4}x^2 - \tfrac{1}{2}x - \tfrac{1}{4} \le \Phi(x) \le -\tfrac{1}{4}x^2 - \tfrac{1}{2}x + \tfrac{3}{4} \qquad \text{for all} \quad x \in\,]-1, +1[,$$
$$x + 2\Phi'(x) + 1 > 0 \qquad \text{for all} \quad x \in\,]-1, +1[,$$
$$1 + 2\Phi''(x) > 0, \qquad \text{for all} \quad x \in\,]-1, +1[,$$
$$\Phi(-1) = \Phi(1) = 0, \qquad \Phi'(-1) = \Phi'(1) = 0$$

and (15) *is satisfied.*

The next result is a consequence of Theorem 1.

Corollary 3. *All solutions of Eq.* (1) *in the class $\mathcal{F}$ can be obtained by the following way. Let us take an arbitrary function $h: [0, \tfrac{1}{2}] \to [0, 1]$ satisfying*

$$h(0) = h'_+(0) = 0, \qquad h'_-(\tfrac{1}{2}) = 1,$$
$$h''(x) > 0, \qquad \text{for all} \quad x \in\,]0, \tfrac{1}{2}].$$

We put

$$f(x) = \begin{cases} h(x) & \text{for} \quad x \in [0, \tfrac{1}{2}], \\ h(1-x) + 2x - 1 & \text{for} \quad x \in\,]\tfrac{1}{2}, 1]. \end{cases}$$

Theorem 6. *Let $f: \mathbf{R} \to \mathbf{R}$ be a differentiable function, for which there exists $f''(0)$. Moreover, let $g: \mathbf{R} \to \mathbf{R}$ be a differentiable function, for which $g'(x) \ne 0$, $x \in \mathbf{R}$ and g' is continuous at 0. If the functions f, g satisfy the following system of functional and differential equations*

$$f(x) = f(1-x) + g(x), \tag{16}$$
$$[g'(x)]^2 \cdot f'(x^2) = g'(x^2) \cdot [f'(x)]^2, \tag{17}$$

then $f(x) = \int x g'(x)\, dx$.

Proof. By Eq. (16) we get

$$f'(x) + f'(1-x) = g'(x) \tag{18}$$

and by Eq. (17) we have $f'(0) = 0$ or $f'(0) = g'(0)$.

Let us define $h(x) := f'(x)/g'(x)$. Since (16) implies $g'(x) = g'(1-x)$, then by (18) we have

$$h(x) + h(1-x) = 1 \tag{19}$$

and $h(0) = 0$ or $h(0) = 1$.

One can observe easily that Eq. (17) implies

$$h(x^2) = h^2(x), \tag{20}$$

therefore $h^2(-x) = h^2(x)$, so

$$h(x) = -h(-x) \quad \text{or} \quad h(x) = h(-x), \qquad \text{for all} \quad x \in \mathbf{R}.$$

Putting $x = \frac{1}{2}$ in Eq. (19) we get $h\left(\frac{1}{2}\right) = \frac{1}{2}$. Putting $x = \frac{1}{2}$ in Eq. (20) we get $h\left(\frac{1}{4}\right) = \frac{1}{4}$ and generally

$$h(1/2^{2^n}) = 1/2^{2^n}, \tag{21}$$

for every $n \in \mathbf{N}$, thus $h(0) = 0$. Remark that since

$$h'_+(0) = \lim_{x \to 0^+} \frac{h(x)}{x} = \lim_{x \to 0^+} \frac{\dfrac{f'(x)}{x}}{g'(x)} = \frac{f''_+(0)}{g'_+(0)},$$

then $h'_+(0)$ exists. Therefore (21) implies that $h(0) = 0$ and $h'_+(0) = 1$.

It is easy to see that there exists $\varepsilon > 0$ such that $h(x) > 0$, for $x \in\,]0, +\varepsilon[$ and $h(x) < 0$, for $x \in\,]-\varepsilon, 0[$, thus we have $h(x) = -h(-x)$, for every $x \in\,]-\varepsilon, +\varepsilon[$.

Indeed, on the contrary we get easily the contradiction with the supposition that there exists $f''(0)$ $(= g'(0) \cdot h'(0))$.

We will prove that the function h is the odd function, so

$$h(x) = -h(-x), \qquad \text{for every} \quad x \in \mathbf{R}. \tag{22}$$

Indeed, on the contrary, let us put $x_0 := \inf\{x \geq \varepsilon \colon h(x) = h(-x)\}$. One can observe easily that $h(x_0) = 0$ necessarily. Putting in Eq. (19) $x := x_0$ we get $h(1 - x_0) = 1$. If $1 - x_0 \leq 0$, then $0 \leq x_0 - 1 \leq x_0$ and $1 = h(1 - x_0) = -h(x_0 - 1) \leq 0$ in view of (19). This implies $1 - x_0 > 0$, so $x_0 < 1$. By Eq. (20) we get $h(x_0^2) = h^2(x_0) = 0$, therefore $h(x_0^{2^n}) = 0$, for every $n \in \mathbf{N}$ and $\lim_{n \to \infty} x_0^{2^n} = 0$. This is the contradiction with $h'_+(0) = 1$, then we have (22).

Replacing x by $x + 1$ in Eq. (19), we get $h(x + 1) + h(-x) = 1$, so $h(x + 1) = -h(-x) + 1$, therefore

$$h(x + 1) = h(x) + 1. \tag{23}$$

By the result of VOLKMANN (see [2]), since the function h satisfies Eqs. (23) and (20), then $h(x) = x$. This implies $f' = x \cdot g'(x)$ and $f(x) = \int x g'(x)\, dx$. $\qquad \square$

Corollary 4. *If* $f: \mathbf{R} \to \mathbf{R}$ *is a differentiable solution of* (1), *for which there exists* $f''(0)$ *and*

$$2f'(x^2) = [f'(x)]^2,$$

then $f(x) = x^2 + c$, *where* $c \in \mathbf{R}$.

3. Convexity. Extremal Points

(1^0) If the group $(G, +)$ is Abelian and divisible by 2, then the set of solutions (f, g) of (2) is evidently convex and it has no extremal points. Indeed, for

$$f_1(x) = \begin{cases} a_1(x) & \text{for} \quad x \in E_1, \\ a_1(\varphi(x)) & \text{for} \quad x \in E_2, \end{cases}$$

where $a_1: E_1 \to G$ and $a_1(x) \neq 0$ and for the solution (f, g) of (2), the pairs $(f + f_1, g)$ and $(f - f_1, g)$ are the solutions of (2),

$$(f, g) = \frac{(f + f_1, g) + (f - f_1, g)}{2}$$

and $(f + f_1, g) \neq (f, g) \neq (f - f_1, g)$.

Analogously the set of solutions of Eq. (2) with f unknown and φ and g given and the set of solutions of (4) are convex and have no extremal points.

(2^0) The set $\mathcal{F} \cap C^2[0, 1]$ and the set $\mathcal{F}_1$ of solutions of (1) of class $C^2[0, 1] \cap \mathcal{F}$ are convex and they have no extremal points. Indeed, let be for $f \in \mathcal{F} \cap C^2[0, 1]$

$$l(x) = \begin{cases} 0 & \text{for} \quad x \in \left[0, \frac{1}{3}\right], \\ \dfrac{m}{288\pi^2} \sin[4\pi(3x - 1)] - \dfrac{m}{24\pi}\left(x - \frac{1}{3}\right) & \text{for} \quad x \in \left]\frac{1}{3}, \frac{1}{2}\right], \end{cases}$$

where $m = \min_{\left[\frac{1}{3}, \frac{1}{2}\right]} f''(x)$, and

$$f_1(x) = \begin{cases} l(x) & \text{for} \quad x \in \left[0, \frac{1}{2}\right], \\ l(1 - x) & \text{for} \quad x \in \left]\frac{1}{2}, 1\right]. \end{cases}$$

Then $f \pm f_1 \in \mathcal{F} \cap C^2[0, 1]$ and

$$\frac{(f + f_1) + (f - f_1)}{2} = f \quad \text{and} \quad f + f_1 \neq f \neq f - f_1. \quad (24)$$

If $f \in \mathcal{F}_1$, then $f \pm f_1 \in \mathcal{F}_1$ too.

Remark 5. The set $\mathcal{F}_0$ of continuous bijections f on $[0,1]$ such that $f(0) = 0$ or equivalently of increasing bijections (respectively: such that $f(0) = 1$ or equivalently of decreasing bijections), is convex and has no extremal points. To be sufficient put, for $f \in \mathcal{F}_0$ and x_0 such that $f(x_0) = \frac{1}{2}$

$$f_1(x) = \begin{cases} \frac{1}{2}f(x) & \text{for } x \in [0, x_0], \\ -\frac{1}{2}f(x) + \frac{1}{2} & \text{for } x \in]x_0, 1], \end{cases}$$

$$\left(\text{respectively}: \quad f_1(x) = \begin{cases} -\frac{1}{2}f(x) + \frac{1}{2} & \text{for } x \in [0, x_0], \\ \frac{1}{2}f(x) & \text{for } x \in]x_0, 1], \end{cases}\right),$$

so as to have $f \pm f_1 \in \mathcal{F}_0$ and (24).

The set of bijections on $[0, 1]$ (continuous too) is not convex $((f + (1 - f))/2 = \frac{1}{2}$ for every bijection f).

Acknowledgement

We are thankful to the referee for all remarks and suggestions.

References

[1] HYERS, D. H. (1941) On the stability of the linear functional equation. Proc. Natl. Acad. Sci. USA **27**: 222–224

[2] VOLKMANN, P. (1983) Caractérisation de la fonction $f(x) = x$ par un systéme de deux équations fonctionnelles. Compt. Rend. Math. Acad. Sci. **5**(1): 27–28

Authors' addresses: Dr. Andrzej Mach, Institute of Mathematics, Jan Kochanowski University, Świętokrzyska 15, 25-406 Kielce, Poland. E-Mail: amach@pu.kielce.pl; Prof. Dr. Dr. h.c. Zenon Moszner, Institute of Mathematics, Akademia Pedagogiczna, Podchorążych 2, 30-084 Kraków, Poland. E-Mail: zmoszner@ap.krakow.pl.

Sitzungsber. Abt. II (2007) 216: 15–32

Sitzungsberichte
Mathematisch-naturwissenschaftliche Klasse Abt. II
Mathematische, Physikalische und Technische Wissenschaften

Remarks on the Stability of a Functional Equation of Quadratic Type

By

Liviu Cădariu and Viorel Radu

(Vorgelegt in der Sitzung der math.-nat. Klasse am 11. Oktober 2007
durch das w. M. Ludwig Reich)

Abstract

Our aim is to present some *generalized stability results of Ulam-Hyers type* for λ-quadratic functional equations of the form $Q_\lambda(F) = 0$, where $\lambda \in \{1, 2\}$, $Q_\lambda(F)$ is given by

$$Q_\lambda(F)(u, v) := F(u + v) + F(u + S(v)) + (\lambda - 1)(F(u - v) + F(u - S(v)))$$
$$- 2^\lambda \left(F(u) + F(v) + F\left(\frac{u + S(u) + v - S(v)}{2} \right) \right.$$
$$+ F\left(\frac{u - S(u) + v + S(v)}{2} \right) \bigg),$$

and the unknown function F is defined on linear spaces $Z = X_1 \times X_2$ and $S = S_{X_1} := P_{X_1} - P_{X_2}$.

Mathematics Subject Classification (2000): 39B62, 39B72, 39B82, 47H10.
Key words: λ-Quadratic functional equation, fixed points, stability.

1. Introduction

Different methods to obtain stability properties for functional equations are known. The *direct method* revealed by HYERS in [17], where the Ulam's problem concerning the stability of homomorphisms was affirmatively answered for Banach spaces, arrived at a very large extent and successful use (see, e.g., [1], [3], [32], [16], [22]).

The interested reader may consult [13], [18], [10], [11] and [19] for details.

On the other hand, in [27], [5] and [6] a *fixed point method* was proposed, by showing that many theorems concerning the stability of Cauchy and Jensen equations are consequences of the fixed point alternative. Subsequently, the method has been successfully used, e.g., in [7], [8], [31], [21], [20] or [24]. It is worth noting that the fixed point method introduces a metrical context and better clarifies the ideas of stability, which is seen to be unambiguously related to fixed points of concrete contractive-type operators on suitable (function) spaces.

We present some *generalized Ulam-Hyers stability results* for functional equations of λ-quadratic type. By using both the direct method and the fixed point method, we slightly extend the results in [25], [26], [9], [16], [22], [28], [29] and [30].

2. Functional Equations of λ-Quadratic Type

Let X_1, X_2 and Y be real linear spaces and consider the Cartesian product $Z := X_1 \times X_2$ together with the linear selfmappings P_{X_1}, P_{X_2} and S, where $P_{X_1}(u) = (u_1, 0)$, $P_{X_2}(u) = (0, u_2)$, $\forall u = (u_1, u_2) \in Z$, and $S = S_{X_1} := P_{X_1} - P_{X_2}$. A function $F: Z \to Y$ is called a λ-*quadratic mapping* ($\lambda \in \{1, 2\}$) iff it satisfies, for all $u, v \in Z$, the following equation:

$$Q_\lambda(F)(u,v) := F(u+v) + F(u+S(v)) + (\lambda-1)(F(u-v) + F(u-S(v)))$$
$$- 2^\lambda \left(F(u) + F(v) + F\left(\frac{u+S(u)+v-S(v)}{2} \right) \right.$$
$$\left. + F\left(\frac{u-S(u)+v+S(v)}{2} \right) \right) = 0. \tag{2.1}$$

Notice that, whenever Z is an inner product space, $F(u) = a \cdot \|P_{X_1} u\|^\lambda \cdot \|P_{X_2} u\|^2$, $\lambda \in \{1, 2\}$, defines a solution of (2.1) for each $a \in \mathbb{R}$.

For $\lambda = 1$ a solution $F: Z \to Y$ is called an *Add Q-type mapping*. If F is a solution of (2.1) for $X_1 = X_2 = X$, then $X \times X \ni u = (x, z) \to f(x, z) := F(u) \in Y$ is an *additive-quadratic mapping on X*, i.e., it verifies the following equation [26]:

$$f(x+y, z+w) + f(x+y, z-w)$$
$$= 2(f(x, z) + f(y, w) + f(x, w) + f(y, z)), \qquad \forall x, y, z, w \in X. \tag{2.2}$$

For $\lambda = 2$, a solution $F: Z \to Y$ is called a *Bi Q-type mapping*. If F verifies (2.1) for $X_1 = X_2 = X$, then $u = (x, z) \to f(x, z) := F(u)$ is a *bi-quadratic mapping*, verifying the following equation [25]:

$$f(x + y, z + w) + f(x + y, z - w) + f(x - y, z + w) + f(x - y, z - w)$$
$$= 4(f(x, z) + f(y, w) + f(x, w) + f(y, z)), \qquad \forall x, y, z, w \in X.$$
$$(2.3)$$

Remark 2.1. Any solution F of (2.1) has the following properties:

(i) $F(0) = 0$; F is an odd mapping for $\lambda = 1$ and an even mapping for $\lambda = 2$;

(ii) $F(2^n \cdot u) = 2^{(\lambda+2)n} \cdot F(u)$, $\forall u \in Z$, $\forall n \in \mathbb{N}$;

(iii) $F \circ S = F$ and $F \circ P_{X_1} = F \circ P_{X_2} = 0$;

(iv) moreover, if $f(x, z) = F(u)$, where $u = (x, z)$, then

(iv.1) for $\lambda = 1, f$ is additive in the first variable and quadratic in the second variable;

(iv.2) for $\lambda = 2, f$ is quadratic in each variable.

We also have the following

Lemma 2.1. *Suppose* $F: Z \to Y$ *is of the form*

$$F(u) = f_2(z)f_1(x), \qquad \forall u = (x, z) \in Z = X_1 \times X_2,$$

with arbitrary nonzero mappings $f_1: X_1 \to Y$ *and* $f_2: X_2 \to \mathbb{R}$. *Then:*

(i) F *is 1-quadratic if* f_1 *is additive and* f_2 *is quadratic;*

(ii) f_1 *is additive if* F *is 1-quadratic and* f_2 *is quadratic;*

(iii) f_2 *is quadratic if* F *is 1-quadratic and* f_1 *is additive;*

(iv) F *is 2-quadratic if and only if* f_1 *and* f_2 *are quadratic.*

2.1. The Generalized Ulam-Hyers Stability for λ-Quadratic Equations

Let us consider a control mapping $\Phi: Z \times Z \to [0, \infty)$ such that, for all $u, v \in Z$,

$$\Psi(u, v) := \sum_{i=0}^{\infty} \frac{\Phi(2^i u, 2^i v)}{2^{(\lambda+2)(i+1)}} < \infty,$$

$$\left(\Psi(u, v) := \sum_{i=1}^{\infty} 2^{(\lambda+2)(i-1)} \Phi\left(\frac{u}{2^i}, \frac{v}{2^i} \right) < \infty, \quad \text{respectively} \right) \quad (2.4)$$

and suppose Y is a Banach space.

Theorem 2.2. *Let* $F: Z \to Y$ *be such a mapping that* $F \circ P_{X_1} + (\lambda - 1)F \circ P_{X_2} = 0$ *and*

$$\|Q_\lambda(F)(u, v)\|_Y \leq \Phi(u, v), \qquad \forall u, v \in Z. \tag{2.5}$$

Then there exists a unique λ*-quadratic mapping* $B: Z \to Y$, *given by*

$$B(u) = \lim_{n \to \infty} \frac{F(2^n u)}{2^{(\lambda+2)n}}, \qquad \left(B(u) = \lim_{n \to \infty} 2^{(\lambda+2)n} \cdot F\left(\frac{u}{2^n}\right) \right), \qquad \forall u \in Z,$$

for which

$$\|F(u) - B(u)\|_Y \leq \Psi(u, u), \qquad \forall u \in Z. \tag{2.6}$$

Proof. We shall use the Hyers' direct method. Letting $u = v$ in (2.5), we obtain

$$\left\| \frac{F(2u)}{2^{\lambda+2}} - F(u) \right\|_Y \leq \frac{\Phi(u, u)}{2^{\lambda+2}}, \qquad \forall u \in Z.$$

In the next step, as usual, one shows that

$$\left\| \frac{F(2^p u)}{2^{(\lambda+2)p}} - \frac{F(2^m u)}{2^{(\lambda+2)m}} \right\|_Y \leq \sum_{i=p}^{m-1} \frac{\Phi(2^i u, 2^i u)}{2^{(\lambda+2)(i+1)}}, \qquad \forall u \in Z, \tag{2.7}$$

for given integers p, m, with $0 \leq p < m$. Using (2.4) and (2.7), $\{F(2^n u)/2^{(\lambda+2)n}\}_{n \geq 0}$ is a Cauchy sequence for any $u \in Z$. Since Y is complete, we can define the mapping $B: Z \to Y$,

$$B(u) = \lim_{n \to \infty} \frac{F(2^n u)}{2^{(\lambda+2)n}}, \qquad \forall u \in Z. \tag{2.8}$$

By using (2.7) for $p = 0$ and $m \to \infty$ we obtain the estimation (2.6).

By (2.5), we have

$$\left\| \frac{F(2^n(u+v))}{2^{(\lambda+2)n}} + \frac{F(2^n(u+S(v)))}{2^{(\lambda+2)n}} \right.$$

$$+ (\lambda - 1)\left(\frac{F(2^n(u-v))}{2^{(\lambda+2)n}} + \frac{F(2^n(u-S(v)))}{2^{(\lambda+2)n}} \right)$$

$$- 2^\lambda \left(\frac{F(2^n(u))}{2^{(\lambda+2)n}} + \frac{F(2^n(v))}{2^{(\lambda+2)n}} + \frac{1}{2^{(\lambda+2)n}} F\left(2^n \left(\frac{u + S(u) + v - S(v)}{2} \right) \right) \right.$$

$$\left. \left. + \frac{1}{2^{(\lambda+2)n}} F\left(2^n \left(\frac{u - S(u) + v + S(v)}{2} \right) \right) \right) \right) \right\|_Y \leq \frac{\Phi(2^n u, 2^n v)}{2^{(\lambda+2)n}},$$

for all $u, v \in Z$. Using (2.4), (2.8) and letting $n \to \infty$, we immediately see that B is a λ-quadratic mapping.

Let B_1 be a λ-quadratic mapping which satisfies (2.6). Then

$$\|B(u) - B_1(u)\|_Y \leq \left\|\frac{B(2^n u)}{2^{(\lambda+2)n}} - \frac{F(2^n u)}{2^{(\lambda+2)n}}\right\|_Y + \left\|\frac{F(2^n u)}{2^{(\lambda+2)n}} - \frac{B_1(2^n u)}{2^{(\lambda+2)n}}\right\|_Y$$

$$\leq 2 \cdot \sum_{k=n}^{\infty} \frac{\Phi(2^k u, 2^k u)}{2^{(\lambda+2)(k+1)}} \longrightarrow 0, \qquad \text{for} \quad n \to \infty.$$

Hence the uniqueness claim for B holds true. $\qquad\qquad\square$

Let us consider a mapping $\varphi: X \times X \times X \times X \to [0, \infty)$ such that $\forall x, y, z, w \in X$,

$$\psi(x, z, y, w) := \sum_{i=0}^{\infty} \frac{\varphi(2^i x, 2^i z, 2^i y, 2^i w)}{2^{(\lambda+2)(i+1)}} < \infty,$$

$$\left(\psi(x, z, y, w) := \sum_{i=1}^{\infty} 2^{(\lambda+2)(i-1)} \varphi\left(\frac{x}{2^i}, \frac{y}{2^i}, \frac{z}{2^i}, \frac{w}{2^i}\right) < \infty, \quad \text{respectively}\right).$$

As a direct consequence of Theorem 2.2, for $\lambda = 1/\lambda = 2$, we obtain:

Corollary 2.3. *Suppose that X is a real linear space, Y is a real Banach space and let $f: X \times X \to Y$ be a mapping such that*

$$\|f(x + y, z + w) + f(x + y, z - w) + (\lambda - 1)(f(x - y, z + w)$$

$$+ f(x - y, z - w)) - 2^\lambda(f(x, z) + f(y, w) + f(x, w) + f(y, z))\|_Y$$

$$\leq \varphi(x, z, y, w),$$

and let $f(x, 0) + (\lambda - 1) \cdot f(0, z) = 0$, for all $x, y, z, w \in X$. Then there exists a unique additive-quadratic/bi-quadratic mapping $b: X \times X \to Y$, given by

$$b(x, z) = \lim_{n\to\infty} \frac{f(2^n x, 2^n z)}{2^{(\lambda+2)n}}, \qquad \left(b(x, z) = \lim_{n\to\infty} 2^{(\lambda+2)n} \cdot f\left(\frac{x}{2^n}, \frac{z}{2^n}\right)\right),$$

$$\forall x, z \in X,$$

such that

$$\|f(x, z) - b(x, z)\|_Y \leq \psi(x, z, x, z), \qquad \forall x, z \in X. \qquad (2.9)$$

Proof. Let us consider $X_1 = X_2 = X$, $u, v \in X \times X$, $u = (x, z)$, $v = (y, w)$, $F(u) = f(x, z)$, and $\Phi(u, v) = \varphi(x, z, y, w)$. Since $\Psi(u, v) = \psi(x, z, y, w) < \infty$, then we can apply Theorem 2.2. Clearly, the mapping b, defined by $b(x, z) = B(u)$ is additive-quadratic/bi-quadratic and verifies (2.6). $\qquad\square$

For $\lambda = 1$ in the above Corollary, we obtain the stability result in ([26], Theorem 7) and, for $\lambda = 2$, that in ([25], Theorem 7).

2.2. Stability Results of Aoki-Rassias Type

For particular forms of the mapping Φ in (2.4), we can obtain interesting consequences. We identify stability properties with unbounded control conditions invoking sums (AOKI [1]) and products (RASSIAS [28–30]) of powers of norms.

Let X_1, X_2 and Y be real linear spaces. Suppose that $Z := X_1 \times X_2$ is endowed with a norm $\|u\|_Z$ and that Y is a real Banach space.

Corollary 2.4. *Let* $F: Z \to Y$ *be a mapping such that*

$$\|Q_\lambda(F)(u,v)\|_Y \leq \varepsilon(\|u\|_Z^p + \|v\|_Z^q), \qquad \forall u, v \in Z,$$

where $p, q \in [0, \lambda + 2)$ *or* $p, q \in (\lambda + 2, \infty)$ *and* $\varepsilon \geq 0$ *are fixed. If* $F \circ P_{X_1} = 0$ *and* $(\lambda - 1)F \circ P_{X_2} = 0$, *then there exists a unique* λ-*quadratic mapping* $B: Z \to Y$, *such that*

$$\|F(u) - B(u)\|_Y \leq \frac{\varepsilon}{|2^{\lambda+2} - 2^p|} \cdot \|u\|_Z^p + \frac{\varepsilon}{|2^{\lambda+2} - 2^q|} \cdot \|u\|_Z^q, \qquad \forall u \in Z.$$

Proof. Consider the mapping $\Phi: Z \times Z \to [0, \infty)$, $\Phi(u,v) = \varepsilon(\|u\|_Z^p + \|v\|_Z^q)$, where $p, q \in [0, \lambda + 2)$ or $p, q \in (\lambda + 2, \infty)$ and $\varepsilon \geq 0$. Then (see (2.4)),

$$\Psi(u,v) = \varepsilon \cdot \frac{\|u\|_Z^p}{|2^{\lambda+2} - 2^p|} + \varepsilon \cdot \frac{\|v\|_Z^q}{|2^{\lambda+2} - 2^q|} < \infty, \qquad \forall u, v \in Z,$$

and the conclusion follows directly from Theorem 2.2.   □

Now, suppose that $X_1 = X_2 = X$, where X is a real normed space, and consider the function $X \times X \ni u = (x, z) \to F(u) = f(x, z)$, where f is mapping $X \times X$ into the real Banach space Y. Although the functions of the form $u \to \|u\| := (\|x\|^r + \|z\|^s)^{1/t}$ may not be norms, the above proofs work as well, and we obtain the following stability properties for λ-quadratic equations:

Corollary 2.5. *Let* $f: X \times X \to Y$ *be a mapping such that*

$$\|f(x+y, z+w) + f(x+y, z-w) + (\lambda - 1)(f(x-y, z+w)$$
$$+ f(x-y, z-w)) - 2^\lambda(f(x,z) + f(y,w) + f(x,w) + f(y,z))\|_Y$$
$$\leq \varepsilon(\|x\|_X^p + \|y\|_X^p + \|z\|_X^q + \|w\|_X^q),$$

for all $x, y, z, w \in X$ *and for some fixed* ε, p, q, *with* $p, q \in [0, 2 + \lambda)$ *or* $p, q \in (\lambda + 2, \infty)$ *and* $\varepsilon \geq 0$. *If* $f(x, 0) = 0$ *and* $(\lambda - 1)f(0, y) = 0$, *for all* $x, y \in X$, *then there exists a unique additive-quadratic/bi-quadratic mapping* $b: X \times X \to Y$, *such that*

$$\|f(x,z) - b(x,z)\|_Y \leq \frac{2\varepsilon}{|2^{\lambda+2} - 2^p|} \cdot \|x\|_X^p + \frac{2\varepsilon}{|2^{\lambda+2} - 2^q|} \cdot \|z\|_X^q, \quad \forall x, z \in X$$

for all $x, z \in X$.

Furthermore, by using the means inequality or directly, two interesting results of RASSIAS type can be obtained for products:

Corollary 2.6. *Let* $F: Z \to Y$ *be a mapping such that*

$$\|Q_\lambda(F)(u,v)\|_Y \leq \varepsilon \cdot \|u\|_Z^p \cdot \|v\|_Z^q, \quad \forall u, v \in Z,$$

where $\varepsilon, p, q \geq 0$ *are fixed and* $p + q \neq \lambda + 2$. *If* $F \circ P_{X_1} = 0$ *and* $(\lambda - 1)F \circ P_{X_2} = 0$, *then there exists a unique* λ-*quadratic mapping* $B: Z \to Y$, *such that*

$$\|F(u) - B(u)\|_Y \leq \frac{\varepsilon}{|2^{\lambda+2} - 2^{p+q}|} \cdot \|u\|_Z^{p+q}, \quad \forall u \in Z.$$

Proof. Consider the mapping $\Phi: Z \times Z \to [0, \infty)$, $\Phi(u,v) = \varepsilon \cdot \|u\|_Z^p \cdot \|v\|_Z^q$, where $\varepsilon, p, q \geq 0$ are fixed and $p + q \neq \lambda + 2$. Then (see (2.4))

$$\Psi(u,v) = \varepsilon \cdot \frac{\|u\|_Z^p \cdot \|v\|_Z^q}{|2^{\lambda+2} - 2^{p+q}|} < \infty, \quad \forall u, v \in Z,$$

so that we can apply Theorem 2.2. $\qquad \square$

Corollary 2.7. *Let* $f: X \times X \to Y$ *be a mapping such that*

$$\|f(x+y, z+w) + f(x+y, z-w) + (\lambda - 1)(f(x-y, z+w)$$
$$+ f(x-y, z-w)) - 2^\lambda(f(x,z) + f(y,w) + f(x,w) + f(y,z))\|_Y$$
$$\leq \varepsilon \cdot (\|x\|_X^p + \|z\|_X^p) \cdot (\|y\|_X^q + \|w\|_X^q),$$

for all $x, y, z, w \in X$ *and for some fixed* $\varepsilon, p, q \geq 0$, *with* $p + q \neq \lambda + 2$. *If* $f(x, 0) = 0$ *and* $(\lambda - 1)f(0, y) = 0$, *for all* $x \in X$, *then there exists a unique additive-quadratic/bi-quadratic mapping* $b: X \times X \to Y$, *such that*

$$\|f(x,z) - b(x,z)\|_Y \leq \frac{\varepsilon}{|2^{\lambda+2} - 2^{p+q}|} \cdot (\|x\|_X^p + \|z\|_X^p) \cdot (\|x\|_X^q + \|z\|_X^q),$$

$$\forall x, z \in X.$$

2.3. Applications to Additive Equations
and to Quadratic Equations

A function $h: X \to Y$, between linear spaces, is called *a mapping of* λ-*order*, $\lambda \in \{1, 2\}$, if it satisfies the following equation:

$$h(x+y) + (\lambda - 1)h(x-y) = 2^{\lambda-1}(h(x) + h(y)), \quad \forall x, y \in X. \quad (2.10)_\lambda$$

Obviously, a mapping of 1-order is an additive mapping and a mapping of 2-order is a quadratic mapping.

For the sake of convenience, we recall the following generalized Ulam-Hyers stability properties of the additive and quadratic functional equations. Let X be a real normed vector space, Y a real Banach space and $\bar{\varphi}: X \times X \to [0, \infty)$ a given mapping.

$\mathbf{A_1}$ ([16], Theorem; see also [12]): *If $\bar{\varphi}$ verifies the condition*

$$\bar{\phi}_1(x, y) := \sum_{i=0}^{\infty} \frac{\bar{\varphi}(2^i x, 2^i y)}{2^{i+1}} < \infty, \qquad \text{for all} \quad x, y \in X \quad (2.11)_1$$

and the mapping $\bar{f}: X \to Y$ satisfies the relation

$$\|\bar{f}(x+y) - \bar{f}(x) - \bar{f}(y)\|_Y \leq \bar{\varphi}(x, y), \quad \text{for all} \quad x, y \in X, \quad (2.12)_1$$

then there exists a unique additive mapping $\bar{a}_1: X \to Y$ which satisfies the inequality

$$\|\bar{f}(x) - \bar{a}_1(x)\|_Y \leq \bar{\phi}_1(x, x), \qquad \text{for all} \quad x \in X. \quad (2.13)_1$$

$\mathbf{A_2}$ ([22], Theorem 2.2): *If $\bar{\varphi}$ verifies the condition*

$$\bar{\phi}_2(x, y) := \sum_{i=0}^{\infty} \frac{\bar{\varphi}(2^i x, 2^i y)}{2^{2(i+1)}} < \infty, \qquad \text{for all} \quad x, y \in X \quad (2.11)_2$$

and the mapping $\bar{f}: X \to Y$, with $\bar{f}(0) = 0$, satisfies the relation

$$\|\bar{f}(x+y) + \bar{f}(x-y) - 2\bar{f}(x) - 2\bar{f}(y)\|_Y \leq \bar{\varphi}(x, y),$$
$$\text{for all} \quad x, y \in X, \quad (2.12)_2$$

then there exists a unique quadratic mapping $\bar{a}_2: X \to Y$ which satisfies the inequality

$$\|\bar{f}(x) - \bar{a}_2(x)\|_Y \leq \bar{\phi}_2(x, x), \qquad \text{for all} \quad x \in X. \quad (2.13)_2$$

As a matter of fact, we can show that the above results are consequences of our Theorem 2.2. Namely, we have

Application 1. The stability of Eq. (2.1) implies the generalized Ulam-Hyers stability of the λ-order equation (2.10)$_\lambda$.

Indeed, let $X, Y, \bar{\varphi}: X \times X \to [0, \infty)$ and $\bar{f}: X \to Y$ be as in $\mathbf{A}_\lambda$, $\lambda \in \{1, 2\}$. We take $X_1 = X$ and consider a linear space X_2 such that there exist a quadratic function $\bar{h}: X_2 \to \mathbb{R}$, with $\bar{h}(0) = 0$ and an element $z_0 \in X_2$, such that $\bar{h}(z_0) \neq 0$. (In inner product spaces such a function is, e.g., $z \to \|z\|^2$.) If we set, for $u = (x, z), v = (y, w) \in X \times X_2$,

$$\Phi(u, v) = \Phi(x, z, y, w) = 2|\bar{h}(z) + \bar{h}(w)| \cdot \bar{\varphi}(x, y)$$

and

$$F(u) = F(x, z) = \bar{h}(z) \cdot \bar{f}(x),$$

then, by using the properties of the quadratic mapping and the relations (2.11)$_\lambda$, for $\lambda \in \{1, 2\}$, we easily get

$$\Psi(u, v) = \frac{1}{2}|\bar{h}(z) + \bar{h}(w)| \sum_{i=0}^{\infty} \frac{\bar{\varphi}(2^i x, 2^i y)}{2^{\lambda(i+1)}} < \infty,$$

for all $u, v \in X \times X_2$. At the same time, by (2.12)$_\lambda$,

$$\|Q_\lambda(F)(u, v)\|_Y = 2|\bar{h}(z) + \bar{h}(w)| \cdot \|\bar{f}(x + y) + (\lambda - 1)\bar{f}(x - y)$$
$$- 2^{\lambda-1}(\bar{f}(x) + \bar{f}(y))\|_Y \leq 2|\bar{h}(z) + \bar{h}(w)| \cdot \bar{\varphi}(x, y)$$
$$= \Phi(u, v), \qquad \forall u, v \in X \times X_2.$$

Therefore, by Theorem 2.2, there exists a unique mapping of λ-quadratic type, $B: X \times X_2 \to Y$, such that $\|F(u) - B(u)\|_Y \leq \Psi(u, u)$ and

$$B(u) = \lim_{n \to \infty} \frac{F(2^n u)}{2^{n(\lambda+2)}} = \lim_{n \to \infty} \frac{\bar{h}(2^n z)}{2^{2n}} \cdot \frac{\bar{f}(2^n x)}{2^{\lambda n}} = \lim_{n \to \infty} \bar{h}(z) \cdot \frac{\bar{f}(2^n x)}{2^{\lambda n}},$$
$$\forall u = (x, z) \in X \times X_2.$$

We know that $\bar{h}(z_0) \neq 0$. Therefore the limit

$$\bar{a}_\lambda(x) := \lim_{n \to \infty} \frac{\bar{f}(2^n x)}{2^{\lambda n}}$$

exists for every $x \in X$ and, moreover, $B(u) = \bar{h}(z) \cdot \bar{a}_\lambda(x), \forall u = (x, z) \in Z$. Since $\|\bar{h}(z)\bar{f}(x) - B(u)\|_Y \leq \bar{h}(z) \cdot \bar{\phi}_\lambda(x, x), \forall u = (x, z) \in X \times X_2$, then the estimation (2.13)$_\lambda$ is easily seen to hold.

By Lemma 2.1, $\bar{a}_1$ is additive and $\bar{a}_2$ is quadratic. If a mapping of λ-order $\bar{c}_\lambda$ satisfies (2.13)$_\lambda$, then $(x, z) \to \bar{h}(z)\bar{c}_\lambda(x)$ is of λ-quadratic type (again by Lemma 2.1) and has to coincide with B, that is

$\bar{h}(z)\bar{a}_\lambda(x) = \bar{h}(z)\bar{c}_\lambda(x)$, for all $u = (x, z) \in X \times X_2$. Since $\bar{h}$ is nonzero, then $\bar{a}_\lambda(x) = \bar{c}_\lambda(x)$, for all $x \in X$. Hence $\bar{a}_\lambda$ is unique.

Remark 2.2. As in the proof of Application 1 for an *additive* function $\bar{h}: X_2 \to \mathbb{R}$, we can also show, by using Theorem 2.2, that the stability of Eq. (2.1) for $\lambda = 1$ implies the generalized Ulam-Hyers stability of the quadratic equation $(2.10)_2$.

As very particular cases, we obtain the results in AOKI [1] and RASSIAS [28] for additive equations:

Application 2. Let $\bar{f}: X \to Y$ be a mapping such that

$$\|\bar{f}(x + y) - \bar{f}(x) - \bar{f}(y)\|_Y \leq \varepsilon(\|x\|_X^p + \|y\|_X^p), \qquad \text{for all} \quad x, y \in X,$$

and for any fixed $\varepsilon, p \geq 0$, with $p \neq 1$. If $\bar{f}(0) = 0$, then there exists a unique additive mapping $\bar{a}_1: X \to Y$ which satisfies the estimation

$$\|\bar{f}(x) - \bar{a}_1(x)\|_Y \leq \frac{2\varepsilon}{|2 - 2^p|} \cdot \|x\|_X^p, \qquad \text{for all} \quad x \in X.$$

Indeed, let $\bar{h}: \mathbb{R} \to \mathbb{R}, \bar{h}(z) = z^2$ and $\bar{f}: X \to Y$, where X is a normed space and Y a Banach space. We apply Theorem 2.2 for $\lambda = 1$, $X_1 = X$, $X_2 = \mathbb{R}$, $u, v \in X \times \mathbb{R}$, with $u = (x, z)$, $v = (y, w)$ and the mappings

$$F(u) = F(x, z) = z^2 \cdot \bar{f}(x),$$

$$\Phi(u, v) = \Phi(x, z, y, w) = 2(z^2 + w^2) \cdot \varepsilon(\|x\|_X^p + \|y\|_X^p),$$

to obtain the existence of a unique additive mapping $\bar{a}_1$ and the required estimation.

Application 3. Let $\bar{f}: X \to Y$ be a mapping such that

$$\|\bar{f}(x + y) - \bar{f}(x) - \bar{f}(y)\|_Y \leq \theta(\|x\|_X^{p/2} \cdot \|y\|_X^{p/2}), \qquad \text{for all} \quad x, y \in X,$$

and for any fixed $\theta, p \geq 0$, with $p < 1$. If $\bar{f}(0) = 0$, then there exists a unique additive mapping $\bar{a}_1: X \to Y$ which satisfies the estimation

$$\|\bar{f}(x) - \bar{a}_1(x)\|_Y \leq \frac{\theta}{2 - 2^p} \cdot \|x\|_X^p, \qquad \text{for all} \quad x \in X.$$

Indeed, one can use *either* the mappings

$$F(u) = F(x, z) = z^2 \cdot \bar{f}(x),$$

and

$$\Phi(u, v) = \Phi(x, z, y, w) = 2(z^2 + w^2) \cdot \theta \cdot \|x\|_X^{p/2} \cdot \|y\|_X^{p/2},$$

or the means inequality:

$$\theta\big(\|x\|_X^{p/2} \cdot \|y\|_X^{p/2}\big) \leq \frac{\theta}{2}\big(\|x\|_X^p + \|y\|_X^p\big)$$

in the preceding corollary.

In particular, we obtain also a stability property of AOKI type for quadratic equations ([9]):

Application 4. Let $\bar{f}$ be a mapping from a real linear space X into a real Banach space Y, such that

$$\|\bar{f}(z+w) + \bar{f}(z-w) - 2\bar{f}(z) - 2\bar{f}(w)\|_Y \leq \varepsilon\big(\|z\|_X^p + \|w\|_X^p\big),$$

$$\text{for all} \quad z, w \in X,$$

and for some fixed $\varepsilon, p \geq 0$, with $p \neq 2$. If $\bar{f}(0) = 0$, then there exists a unique quadratic mapping $\bar{a}_2 \colon X \to Y$ which satisfies the estimation

$$\|\bar{f}(z) - \bar{a}_2(z)\|_Y \leq \frac{2\varepsilon}{|2^2 - 2^p|} \cdot \|z\|_X^p, \qquad \text{for all} \quad z \in X.$$

For the *proof*, let $\bar{h} \colon \mathbb{R} \to \mathbb{R}, \bar{h}(x) = x$. We apply Theorem 2.2 for $\lambda = 1$, $X_1 = \mathbb{R}$, $X_2 = X$, $u, v \in \mathbb{R} \times X$, with $u = (x, z)$, $v = (y, w)$ and the mappings $F(u) = F(x, z) = x \cdot \bar{f}(z)$, $\Phi(u, v) = \Phi(x, z, y, w) = |x + y| \cdot \varepsilon\big(\|z\|_X^p + \|w\|_X^p\big)$, to obtain the existence of a unique quadratic mapping $\bar{a}_2$ and the required estimation.

Similarly, by choosing $F(u) = F(x, z) = x \cdot \bar{f}(z)$ and $\Phi(u, v) = \Phi(x, z, y, w) = |x + y| \cdot \varepsilon \cdot \|z\|_X^p \cdot \|w\|_X^q$, we obtain a stability of RASSIAS type [30]:

Application 5. Let $\bar{f}$ be a mapping from a real linear space X into a real Banach space Y such that $\bar{f}(0) = 0$ and

$$\|\bar{f}(z+w) + \bar{f}(z-w) - 2\bar{f}(z) - 2\bar{f}(w)\|_Y \leq \varepsilon \cdot \|z\|_X^p \cdot \|w\|_X^q, \quad \forall z, w \in X,$$

for some fixed $\varepsilon, p, q \geq 0$, with $p + q \neq 2$. Then there exists a unique quadratic mapping $\bar{a}_2 \colon X \to Y$ which satisfies the estimation

$$\|\bar{f}(z) - \bar{a}_2(z)\|_Y \leq \frac{\varepsilon}{|2^2 - 2^{p+q}|} \cdot \|z\|_X^{p+q}, \qquad \forall z \in X.$$

3. A Second Stability Result by the Fixed Point Method

We will show that Corollary 2.4 and Corollary 2.6 can be essentially extended by using *a fixed point method*. The method is seen plainly related to some fixed point of a concrete operator. Specifically, our control conditions are perceived to be responsible for three fundamental facts: Actually, they ensure

   1) the *contraction property* of a Schröder type operator J and

   2) the first two successive approximations, f and Jf, to be at a *finite distance*.

And, moreover, they force

   3) the fixed point function of J to be a *solution of the initial equation*.

   Firstly, we prove an auxiliary result of stability for the following equation in a single variable

$$w \circ g \circ \eta = g.$$

Let us consider a Lipschitzian function $w \colon Y \to Y$, with the Lipschitz constant L_w, and the mappings $f \colon G \to Y$, $\eta \colon G \to G$, where G is a nonempty set and Y is a Banach space.

Lemma 3.1. *Suppose that the mapping f satisfies an inequality of the form*

$$\| (w \circ f \circ \eta)(x) - f(x) \|_Y \leq \psi(x), \qquad \forall x \in G, \qquad (C_\psi)$$

where $\psi \colon G \to [0, \infty)$. If there exists $L < 1$ such that the mapping ψ has the property

$$L_w \cdot (\psi \circ \eta)(x) \leq L\psi(x), \qquad \forall x \in G, \qquad (H_\psi)$$

then there exists a unique mapping $c \colon G \to Y$,

$$c(x) := \lim_{n \to \infty} (w^n \circ c \circ \eta^n)(x), \qquad \forall x \in G,$$

which satisfies the equation

$$(w \circ c \circ \eta)(x) = c(x), \qquad \forall x \in G$$

and the inequality

$$\| f(x) - c(x) \|_Y \leq \frac{\psi(x)}{1 - L}, \qquad \forall x \in G. \qquad (Est_\psi)$$

Proof. Let us consider the set $\mathcal{E} := \{ g \colon G \to Y \}$ and introduce a *complete generalized metric* on $\mathcal{E}$ (as usual, $\inf \emptyset = \infty$):

$$d(g, h) = d_\psi(g, h) = \inf\{ K \in \mathbb{R}_+, \| g(x) - h(x) \|_Y \leq K\psi(x), \forall x \in G \}. \qquad (GM_\psi)$$

Now, define the mapping

$$J \colon \mathcal{E} \to \mathcal{E}, \, Jg(x) := (w \circ g \circ \eta)(x). \qquad (OP)$$

Step I. By using the hypothesis (H_ψ), we show that J *is strictly contractive* on $\mathcal{E}$.

We can write, for any $g, h \in \mathcal{E}$:

$$d(g, h) < K \Longrightarrow \|g(x) - h(x)\|_Y \le K\psi(x), \qquad \forall x \in G.$$

On the other hand,

$$\begin{aligned}
\|Jg(x) - Jh(x)\|_Y &= \|w(g(\eta(x))) - w(h(\eta(x)))\|_Y \\
&\le L_w \cdot \|g(\eta(x)) - h(\eta(x))\|_Y \le L_w \cdot K \cdot \psi(\eta(x)) \\
&\le K \cdot L \cdot \psi(x), \quad \forall x \in G \Longrightarrow d(Jg, Jh) \le LK.
\end{aligned}$$

Therefore, we see that

$$d(Jg, Jh) \le Ld(g, h), \qquad \forall g, h \in \mathcal{E}, \tag{CC_L}$$

that is J is a *strictly contractive* self-mapping of $\mathcal{E}$, with the constant $L < 1$.

Step II. Obviously, $d(f, Jf) < \infty$.

In fact, by using the relation (C_ψ), it results that $d(f, Jf) < 1$.

Step III. We can apply the fixed point alternative (see, e.g., [5]), and we obtain the existence of a mapping $c: G \to Y$ such that:

— c is a fixed point of J, that is

$$(w \circ c \circ \eta)(x) = c(x), \qquad \forall x \in G. \tag{3.1}$$

The mapping c is the unique fixed point of J in the set

$$\mathcal{F} = \{g \in \mathcal{E}, \ d(f, g) < \infty\}.$$

This says that c is the unique mapping with *both* the properties (3.1) and (3.2), where

$$\exists K \in (0, \infty) \ \text{such that} \ \|c(x) - f(x)\|_Y \le K\psi(x), \ \forall x \in G. \tag{3.2}$$

— $d(J^n f, c) \underset{n \to \infty}{\longrightarrow} 0$, which implies the equality

$$c(x) := \lim_{n \to \infty} (w^n \circ c \circ \eta^n)(x), \qquad \forall x \in G. \tag{3.3}$$

— $d(f, c) \le \dfrac{1}{1 - L} d(f, Jf)$, which implies the inequality

$$d(f, c) \le \frac{1}{1 - L},$$

that is (Est_ψ) is seen to be true. $\qquad\qquad\qquad\qquad\qquad \square$

Let X_1, X_2 be linear spaces, $Z := X_1 \times X_2$, Y a Banach space, and consider an arbitrary mapping $\Phi: Z \times Z \to [0, \infty)$.

Theorem 3.2. *Let $F: Z \to Y$ be such a mapping for which $F \circ P_{X_1} + (\lambda - 1)F \circ P_{X_2} = 0$ and suppose that*

$$\|Q_\lambda(F)(u, v)\|_Y \le \Phi(u, v), \qquad \forall u, v \in Z. \tag{2.5}$$

If there exists $L < 1$ such that the mapping

$$u \to \Omega(u) = \Phi\left(\frac{u}{2}, \frac{u}{2}\right)$$

verifies the condition

$$\Omega(u) \le L \cdot 2^{\lambda+2} \cdot \Omega\left(\frac{u}{2}\right), \qquad \forall u \in Z, \tag{H_λ}$$

and the mapping Φ has the property

$$\lim_{n \to \infty} \frac{\Phi(2^n u, 2^n v)}{2^{(\lambda+2)n}} = 0, \qquad \forall u, v \in Z, \tag{H_λ^*}$$

then there exists a unique λ-quadratic mapping $B: Z \to Y$, such that

$$\|F(u) - B(u)\|_Y \le \frac{L}{1 - L}\Omega(u), \qquad \forall u \in Z. \tag{Est}$$

Proof. If we set $u = v$ in the relation (2.5), then we see that

$$\|F(2u) - 2^{\lambda+2}F(u)\|_Y \le \Omega(2u), \qquad \forall u \in Z.$$

Hence

$$\left\|\frac{F(2u)}{2^{\lambda+2}} - F(u)\right\|_Y \le \frac{\Omega(2u)}{2^{\lambda+2}}, \qquad \forall u \in Z. \tag{3.4}$$

Now we can apply Lemma 3.1, with $w, \eta: Z \to Y$, $\psi: Z \to [0, \infty)$,

$$w(u) := \frac{u}{2^{\lambda+2}}, \qquad \eta(u) := 2u, \qquad \psi(u) := \frac{\Omega(2u)}{2^{\lambda+2}}.$$

Clearly, $L_w = 1/2^{\lambda+2}$ and, by using (3.4) and the hypothesis (H_λ), we obtain that (C_ψ) and (H_ψ) hold.

Then there exists a unique mapping $B: Z \to Y$,

$$B(u) := \lim_{n \to \infty}(w^n \circ B \circ \eta^n)(u) = \lim_{n \to \infty}\frac{F(2^n u)}{2^{(\lambda+2)n}}, \qquad \forall u \in Z, \tag{3.5}$$

which satisfies the following equation

$$(w \circ B \circ \eta)(u) = B(u) \Leftrightarrow B(2u) = 2^{\lambda+2}B(u), \qquad \forall u \in Z$$

and the inequality

$$\|F(u) - B(u)\|_Y \le \frac{\psi(u)}{1 - L} = \frac{\Omega(2u)}{2^{\lambda+2}} \cdot \frac{1}{1 - L} \le \Omega(u)\frac{L}{1 - L}, \qquad \forall u \in Z.$$

The statement that B is a λ-quadratic mapping is easily seen: If we replace u by $2^n u$ and v by $2^n v$ in (2.5), then we obtain

$$\left\| \frac{F(2^n(u + v))}{2^{(\lambda+2)n}} + \frac{F(2^n(u + S(v)))}{2^{(\lambda+2)n}} \right.$$
$$+ (\lambda - 1)\left(\frac{F(2^n(u - v))}{2^{(\lambda+2)n}} + \frac{F(2^n(u - S(v)))}{2^{(\lambda+2)n}} \right) - 2^\lambda \left(\frac{F(2^n(u))}{2^{(\lambda+2)n}} \right.$$
$$+ \frac{F(2^n(v))}{2^{(\lambda+2)n}} + \frac{1}{2^{(\lambda+2)n}} F\left(2^n \left(\frac{u + S(u) + v - S(v)}{2} \right) \right)$$
$$\left. + \frac{1}{2^{(\lambda+2)n}} F\left(2^n \left(\frac{u - S(u) + v + S(v)}{2} \right) \right) \right) \right\|_Y \le \frac{\Phi(2^n u, 2^n v)}{2^{(\lambda+2)n}},$$

for all $u, v \in Z$. By using (3.5) and (H_λ^*) and letting $n \to \infty$, we see that B satisfies (2.1).   □

Example 3.1. If we apply Theorem 3.2 with the mappings $\Phi: Z \times Z \to [0, \infty)$ given by $(u, v) \to \varepsilon(\|u\|_Z^p + \|v\|_Z^q)$ and $(u, v) \to \varepsilon\|u\|_Z^p \cdot \|v\|_Z^q$, then we obtain the stability results in Corollary 2.4 and Corollary 2.6, respectively.

As it is well known (see [15, 18, 9]), GAJDA/CZERWIK showed that the additive/quadratic equation $(2.12)_\lambda$ is *not stable* for $\bar\varphi(x, y)$ of the form $\varepsilon(\|x\|^\lambda + \|y\|^\lambda)$, ε being a given positive constant $(\lambda \in \{1, 2\})$. In fact, it has been proved that there exists a mapping $\bar f_\lambda: \mathbb{R} \to \mathbb{R}$ such that $(2.12)_\lambda$ holds with the above $\bar\varphi$, and there exists *no* additive/quadratic mapping $\bar a$ to verify

$$|\bar f_\lambda(x) - \bar a_\lambda(x)| \le c(\varepsilon)|x|^\lambda, \qquad \text{for all} \quad x \in \mathbb{R}.$$

This suggests the following

Example 3.2. Let $X_1 = X_2 = Y = \mathbb{R}$, with the Euclidean norm, and $\bar h: \mathbb{R} \to \mathbb{R}$ a quadratic function with $\bar h(0) = 0, \bar h(1) = 1$. Then *Eq.* (2.1) *is not stable* for

$$\Phi(u, v) = \Phi(x, z, y, w) = 2\varepsilon \cdot (|x|^\lambda + |y|^\lambda)(\bar h(z) + \bar h(w)). \qquad (3.6)$$

In fact, we can show that there exists an F for which the relation (2.5) holds and there exists *no Add Q/Bi Q*-type mapping $B: X_1 \times X_2 \to Y$

to verify

$$|F(u) - B(u)| \leq c(\varepsilon)\bar{h}(z)|x|^{\lambda}, \qquad \forall u = (x,z) \in X_1 \times X_2. \qquad (3.7)$$

Indeed, for $F(u) = F(x,z) = \bar{h}(z) \cdot \bar{f}_{\lambda}(x)$, and Φ as in (3.6), (2.5) holds. Therefore

$$|f(x+y) + (\lambda - 1)f(x-y) - 2^{\lambda-1}(f(x) + h(y))| \leq \varepsilon(|x|^{\lambda} + |y|^{\lambda}),$$

$$\text{for all} \quad x, y \in X_1.$$

Let us suppose, for a contradiction, that there exists an *Add Q/Bi Q*-type mapping B which verifies (3.7). By Remark 2.1, the mapping $\bar{a}_{\lambda} \colon X_1 \to Y$, $\bar{a}_{\lambda}(x) := B(x,1)$ is a solution for $(2.10)_{\lambda}$. The estimation (3.7) gives us

$$|\bar{f}_{\lambda}(x) - \bar{a}_{\lambda}(x)| \leq c(\varepsilon)|x|^{\lambda}, \qquad \forall x \in X_1,$$

in contradiction with the above result of GAJDA/CZERWIK.

References

[1] AOKI, T. (1950) On the stability of the linear transformation in Banach spaces. J. Math. Soc. Japan **2**: 64–66

[2] BAKER, J. A. (1991) The stability of certain functional equations. Proc. Amer. Math. Soc. **3**: 729–732

[3] BOURGIN, D. G. (1951) Classes of transformations and bordering transformations. Bull. Amer. Math. Soc. **57**: 223–237

[4] CĂDARIU, L. (2002) A general theorem of stability for the Cauchy's equation. Bull. Şt. Univ. Politehnica Timişoara, Seria Matematică-Fizică **47**(61) (no. 2): 14–28

[5] CĂDARIU, L., RADU, V. (2003) Fixed points and the stability of Jensen's functional equation. J. Inequal. Pure and Appl. Math. **4**(1): Art. 4 (http://jipam.vu.edu.au)

[6] CĂDARIU, L., RADU, V. (2004) On the stability of Cauchy's functional equation: A fixed points approach. In: SOUSA RAMOS, J., GRONAU, D., MIRA, C., REICH, L., SHARKOVSKY, A. (eds.) Iteration Theory (ECIT '02), Proceedings of the European Conference of Iteration Theory, Evora, Portugal, September 1–7, 2002, Grazer Math. Ber. **346**: 43–52

[7] CĂDARIU, L. (2005) Fixed points in generalized metric spaces and the stability of a quartic functional equation. Bull. Şt. Univ. "Politehnica" Timişoara, Seria Matematică-Fizică **50**(64) (no. 2): 25–34

[8] CĂDARIU, L., RADU, V. (2007) Fixed points in generalized metric spaces and the stability of a cubic functional equation. In: CHO, Y. J., KIM, J. K., KANG, S. M. (eds.) Fixed Point Theory and Applications, Vol. 7, pp. 53–68. Nova Science Publ., Hauppauge, NY

[9] CZERWIK, S. (1992) On the stability of quadratic mapping in normed spaces. Abh. Math. Sem. Univ. Hamburg **62**: 59–64

[10] CZERWIK, S. (2002) Functional Equations and Inequalities in Several Variables. World Scientific, Singapore

[11] DAROCZY, Z., PALES, ZS. (eds.) (2002) Functional Equations – Results and Advances. Kluwer, Dordrecht

[12] FORTI, G. L. (1980) An existence and stability theorem for a class of functional equations. Stochastica **4**: 23–30

[13] FORTI, G. L. (1995) Hyers-Ulam stability of functional equations in several variables. Aequationes Math. **50**: 143–190

[14] FORTI, G. L. (2004) Comments on the core of the direct method for proving Hyers-Ulam stability of functional equations. J. Math. Anal. Appl. **295**(1): 127–133

[15] GAJDA, Z. (1991) On stability of additive mappings. Internat. J. Math. Math. Sci. **14**: 431–434

[16] GĂVRUȚA, P. (1994) A generalization of the Hyers-Ulam-Rassias stability of approximately additive mappings. J. Math. Anal. Appl. **184**: 431–436

[17] HYERS, D. H. (1941) On the stability of the linear functional equation. Proc. Natl. Acad. Sci. USA **27**: 222–224

[18] HYERS, D. H., ISAC, G., RASSIAS, TH. M. (1998) Stability of Functional Equations in Several Variables. Birkhäuser, Basel

[19] JUNG, S. M. (2002) Hyers-Ulam-Rassias Stability of Functional Equations in Mathematical Analysis. Hadronic Press, Palm Harbor, FL

[20] JUNG, S.-M., KIM, T.-S. (2006) A fixed point approach to the stability of cubic functional equations. Bol. Soc. Mat. Mexicana **3**(12): 51–57

[21] JUNG, S.-M. (2007) A fixed point approach to the stability of isometries. J. Math. Anal. Appl. **329**: 879–890

[22] KIM, G. H. (2001) On the stability of the quadratic mapping in normed spaces. Internat. J. Math. Math. Sci. **25**(4): 217–229

[23] MARGOLIS, B., DIAZ, J. B. (1968) A fixed point theorem of the alternative for contractions on a generalized complete metric space. Bull. Amer. Math. Soc. **74**: 305–309

[24] MIRZAVAZIRI, M., MOSLEHIAN, M. S. (2006) A fixed point approach to stability of a quadratic equation. Bull. Braz. Math. Soc. **37** (no. 3): 361–376

[25] PARK, W. G., BAE, J. H. (2005) On a bi-quadratic functional equation and its stability. Nonlinear Analysis **62**(4): 643–654

[26] PARK, W. G., BAE, J. H., CHUNG, B.-H. (2005) On an additive-quadratic functional equation and its stability. J. Appl. Math. Comp. **18**(1–2): 563–572

[27] RADU, V. (2003) The fixed point alternative and the stability of functional equations. Fixed Point Theory **4** (no. 1): 91–96

[28] RASSIAS, J. M. (1982) On approximation of approximately linear mappings by linear mappings. J. Funct. Anal. **46** (no. 1): 126–130

[29] RASSIAS, J. M. (1989) Solution of a problem of Ulam. J. Approx. Theory **57** (no. 3): 268–273

[30] RASSIAS, J. M. (1992) On the stability of the Euler-Lagrange functional equation. C. R. Acad. Bulgare Sci. **45** (no. 6): 17–20

[31] RASSIAS, J. M. (2006) Alternative contraction principle and alternative Jensen and Jensen type mappings. Internat. Journal of Applied Math. & Stat. **4** (no. M06): 1–10

[32] RASSIAS, TH. M. (1978) On the stability of the linear mapping in Banach spaces. Proc. Amer. Math. Soc. **72**: 297–300

[33] RUS, I. A. (2001) Generalized Contractions and Applications. Cluj University Press, Cluj-Napoca
[34] ULAM, S. M. (1960) A Collection of Mathematical Problems. Interscience, New York; ULAM, S. M. (1964) Problems in Modern Mathematics. Wiley, New York

Authors' addresses: Dr. Liviu Cădariu, Departamentul de Matematică, Universitatea Politehnica din Timişoara, Piaţa Victoriei 2, 300006 Timişoara, România. E-Mail: liviu.cadariu@mat.upt.ro, lcadariu@yahoo.com; Dr. Viorel Radu, Facultatea de Matematică, Universitatea de Vest din Timişoara, Bv. Vasile Pârvan 4, 300223 Timişoara, România. E-Mail: radu@math.uvt.ro.

Sitzungsber. Abt. II (2007) 216: 33–43

Sitzungsberichte
Mathematisch-naturwissenschaftliche Klasse Abt. II
Mathematische, Physikalische und Technische Wissenschaften

Generalizations of Implication Algebras

By

Ivan Chajda[1] and Helmut Länger

(Vorgelegt in der Sitzung der math.-nat. Klasse am 11. Oktober 2007
durch das w. M. Ludwig Reich)

Abstract

Implication algebras, originally introduced in order to study algebraic properties of the implication operation in Boolean algebras, are generalized and it is shown that these more general algebras are in one-to-one correspondence to semilattices with 1 the principal filters of which are posets with an antitone involution, respectively to commutative directoids with 1 the principal filters of which are posets with a switching involution.

Mathematics Subject Classification (2000): 20N02, 06A12, 06E05, 06C15.
Key words: Implication algebra, orthoimplication algebra, orthomodular implication algebra, strong *I*-algebra, weak *I*-algebra, semilattice, directoid, Boolean algebra, orthomodular lattice, poset, antitone involution, switching involution.

In order to study algebraic properties of the implication operation in Boolean algebras J. C. ABBOTT introduced the notion of an implication algebra (cf. [1]). He showed that these algebras are in one-to-one correspondence to join-semilattices with 1 the principal filters of which are Boolean algebras. These algebras were generalized, e.g. in [2], [8] and [9], where corresponding results were achieved.

Let us mention that also other types of implication in non-classical logic were treated in the literature (see e.g. [4]–[7]). However, these

---

[1] Support of the research of the first author by the Czech Government Research Project MSM 6198959214 is gratefully acknowledged.

can be unified by a more generalized approach which will be presented here.

The aim of this paper is to further generalize the concept of an implication algebra to algebras satisfying weaker conditions. It turns out that by the constructions originally used by J. C. ABBOTT these more general algebras are in one-to-one correspondence to semilattices with 1 the principal filters of which are posets with an antitone involution, respectively to directoids with 1 the principal filters of which are posets with a switching involution.

1. Implication Algebras, Orthoimplication Algebras and Orthomodular Implication Algebras

We start our investigations by repeating the definition of an implication algebra and its connection to certain join-semilattices having an additional structure.

Definition 1.1. Let $\mathcal{A} = (A, \cdot, 1)$ be an algebra of type $(2, 0)$. $\mathcal{A}$ is called an *implication algebra* (cf. [1]) if it satisfies

$$xx = 1,$$

$$(xy)x = x,$$

$$(xy)y = (yx)x$$

and

$$x(yz) = y(xz).$$

Remark 1.2. The nullary operation 1 can be dropped from the family of fundamental operations of an implication algebra since due to the first identity in the definition it is an algebraic constant.

The following theorem was proved in [1]:

Theorem 1.3. *Let $\mathcal{A} = (A, \cdot, 1)$ be an implication algebra. Define*

$$x \vee y := (xy)y \qquad and \qquad x^y := xy$$

for all $x, y \in A$. Then $\mathbf{S}(\mathcal{A}) := (A, \vee, (^x; x \in A), 1)$ is an algebra such that $(A, \vee, 1)$ is a join-semilattice with greatest element 1 and for every $x \in A$ $([x, 1], \leq, ^x)$ is a Boolean algebra. Conversely, let $\mathcal{S} := (S, \vee, (^x; x \in S), 1)$ be an algebra such that $(S, \vee, 1)$ is a join-semilattice with greatest element 1 and for every $x \in S$ $([x, 1], \leq, ^x)$ is a Boolean algebra. Define

$$xy := (x \vee y)^y$$

for all $x, y \in S$. *Then* $\mathbf{A}(\mathcal{S}) := (S, \cdot, 1)$ *is an implication algebra. Moreover,* $\mathbf{A}(\mathbf{S}(\mathcal{A})) = \mathcal{A}$ *and* $\mathbf{S}(\mathbf{A}(\mathcal{S})) = \mathcal{S}$ *for every implication algebra* $\mathcal{A}$ *and every algebra* $\mathcal{S} = (S, \vee, (^x; x \in S), 1)$ *such that* $(S, \vee, 1)$ *is a join-semilattice with greatest element* 1 *and for every* $x \in S$ $([x, 1], \leq, ^x)$ *is a Boolean algebra.*

The notion of an implication algebra was generalized to the notion of an orthoimplication algebra, respectively orthomodular implication algebra, as follows:

Definition 1.4. Let $\mathcal{A} = (A, \cdot, 1)$ be an algebra of type $(2, 0)$. $\mathcal{A}$ is called an *orthoimplication algebra* (cf. [2]) if it satisfies

$$xx = 1,$$

$$(xy)x = x,$$

$$(xy)y = (yx)x$$

and

$$x((yx)z) = xz.$$

Definition 1.5. Let $\mathcal{A} = (A, \cdot, 1)$ be an algebra of type $(2, 0)$. $\mathcal{A}$ is called an *orthomodular implication algebra* (cf. [8] and [9]) if it satisfies

$$xx = 1,$$

$$(xy)x = x,$$

$$(xy)y = (yx)x,$$

$$(((xy)y)z)(xz) = 1$$

and

$$((((((xy)y)z)x)x)z)x)x = (((xy)y)z)z.$$

In [2], [8] and [9] it was proved that orthoimplication algebras, respectively orthomodular implication algebras, correspond to join-semilattices with 1 the principal filters of which are orthomodular lattices satisfying the compatibility condition ($x \leq y \leq z$ implies $z^y = z^x \vee y$) respectively to join-semilattices with 1 the principal filters of which are orthomodular lattices. These correspondences are one-to-one and completely analogous to that proved by J. C. ABBOTT in [1] for implication algebras.

2. *I*-Algebras

We now define a new more general type of implication algebras:

Definition 2.1. Let $\mathcal{A} = (A, \cdot, 1)$ be an algebra of type $(2,0)$. $\mathcal{A}$ is called a *strong I-algebra* if it satisfies

$$1x = x, \tag{S1}$$
$$xx = 1, \tag{S2}$$
$$x(yx) = 1, \tag{S3}$$
$$(xy)y = (yx)x \tag{S4}$$

and

$$(((xy)y)z)(xz) = 1. \tag{S5}$$

Remark 2.2. The class of all strong I-algebras forms a variety.

For the following theorem we need the definition of an antitone involution of a poset.

Definition 2.3. Let $(P, \leq)$ be a poset and $f\colon P \to P$. f is called an *antitone involution* of $(P, \leq)$ if $f(x) \geq f(y)$ whenever both $x, y \in P$ and $x \leq y$ and if $f(f(x)) = x$ for all $x \in P$.

Now the following result can be proved:

Theorem 2.4. *Let* $\mathcal{A} = (A, \cdot, 1)$ *be a strong I-algebra. Define*

$$x \vee y := (xy)y \qquad and \qquad x^y := xy$$

for all $x, y \in A$. *Then* $\mathbf{S}(\mathcal{A}) := (A, \vee, (^x; x \in A), 1)$ *is an algebra such that* $(A, \vee, 1)$ *is a join-semilattice with greatest element* 1 *and for every* $x \in A$ $([x, 1], \leq, ^x)$ *is a poset with an antitone involution where* $\leq$ *denotes the partial order induced by* $\vee$. *Conversely, let* $\mathcal{S} := (S, \vee, (^x; x \in S), 1)$ *be an algebra such that* $(S, \vee, 1)$ *is a join-semilattice with greatest element* 1 *and for every* $x \in S$ $([x, 1], \leq, ^x)$ *is a poset with an antitone involution. Define*

$$xy := (x \vee y)^y$$

for all $x, y \in S$. *Then* $\mathbf{A}(\mathcal{S}) := (S, \cdot, 1)$ *is a strong I-algebra. Moreover,* $\mathbf{A}(\mathbf{S}(\mathcal{A})) = \mathcal{A}$ *and* $\mathbf{S}(\mathbf{A}(\mathcal{S})) = \mathcal{S}$ *for every strong I-algebra* $\mathcal{A}$ *and every algebra* $\mathcal{S} = (S, \vee, (^x; x \in S), 1)$ *such that* $(S, \vee, 1)$ *is a join-semilattice with greatest element* 1 *and for every* $x \in S$ $([x, 1], \leq, ^x)$ *is a poset with an antitone involution.*

Proof. Assume $(A, \cdot, 1)$ to be a strong I-algebra and for all $x, y \in A$ define $x \leq y$ if $xy = 1$, $x \vee y := (xy)y$ and $x^y := xy$. Because of (S2), $\leq$ is reflexive. If $a \leq b$ and $b \leq a$ then

$$a = 1a = (ba)a = (ab)b = 1b = b$$

according to (S1) and (S4) proving antisymmetry of $\leq$. If $a \leq b \leq c$ then

$$ac = 1(ac) = (bc)(ac) = ((1b)c)(ac) = (((ab)b)c)(ac) = 1$$

according to (S1) and (S5), i.e., $a \leq c$. This shows transitivity of $\leq$. Hence $(A, \leq)$ is a poset and according to (S2) and (S3), $a1 = a(aa) = 1$, i.e., $a \leq 1$ which means that 1 is the greatest element of $(A, \leq)$. According to (S4), $\vee$ is commutative. Because of (S1), (S2) and (S5)

$$a(a \vee b) = a((ab)b) = 1(a((ab)b)) = (((ab)b)((ab)b))(a((ab)b)) = 1,$$

i.e., $a \leq a \vee b$. Hence $b \leq b \vee a = a \vee b$. Because of (S1) and (S5), $a \leq b$ implies

$$(bc)(ac) = ((1b)c)(ac) = (((ab)b)c)(ac) = 1,$$

i.e., $bc \leq ac$. Therefore $a, b \leq c$ implies $cb \leq ab$ and hence

$$a \vee b = (ab)b \leq (cb)b = (bc)c = 1c = c$$

according to (S4) and (S1). Hence $a \vee b$ is the supremum of a and b with respect to $\leq$. If $b \in [a, 1]$ then $ab^a = a(ba) = 1$ according to (S3), i.e., $b^a \in [a, 1]$. Hence ^{a} is a unary operation on $[a, 1]$. If $a \leq b \leq c$ then $c^a = ca \leq ba = b^a$, i.e., ^{a} is antitone. If $a \leq b$ then $(b^a)^a = (ba)a = (ab)b = 1b = b$ according to (S4) and (S1) and hence ^{a} is an involution. Therefore $(A, \vee, (^a; a \in A), 1)$ is an algebra such that $(A, \vee, 1)$ is a join-semilattice with greatest element 1 and for every $x \in A$ ($[x, 1], \leq, ^x$) is a poset with an antitone involution with respect to the partial order induced by $\vee$. Moreover, $((ab)b)b = ab \vee b = ab$ since $b \leq ab$ according to $b(ab) = 1$ by (S3).

Conversely, assume $(A, \vee, (^a; a \in A), 1)$ to be an algebra such that $(A, \vee, 1)$ is a join-semilattice with greatest element 1 and for every $x \in A$ ($[x, 1], \leq, ^x$) is a poset with an antitone involution. Moreover, for all $x, y \in A$ define $xy := (x \vee y)^y$.

$$1a = (1 \vee a)^a = 1^a = a, \tag{S1}$$

$$aa = (a \vee a)^a = a^a = 1, \tag{S2}$$

$$a(ba) = (a \vee (b \vee a)^a)^{(b \vee a)^a} = ((b \vee a)^a)^{(b \vee a)^a} = 1, \tag{S3}$$

$$(ab)b = ((a \vee b)^b \vee b)^b = ((a \vee b)^b)^b = a \vee b = b \vee a$$
$$= ((b \vee a)^a)^a = ((b \vee a)^a \vee a)^a = (ba)a, \tag{S4}$$

$$(((ab)b)c)(ac) = ((((a \vee b)^b \vee b)^b \vee c)^c \vee (a \vee c)^c)^{(a \vee c)^c}$$
$$= ((((a \vee b)^b)^b \vee c)^c \vee (a \vee c)^c)^{(a \vee c)^c}$$
$$= ((a \vee b \vee c)^c \vee (a \vee c)^c)^{(a \vee c)^c}$$
$$= ((a \vee c)^c)^{(a \vee c)^c} = 1. \tag{S5}$$

Therefore $(A, \cdot, 1)$ is a strong I-algebra. Moreover, $((a \vee b)^b \vee b)^b = ((a \vee b)^b)^b = a \vee b$ and if $a \leq b$ then $(b \vee a)^a = b^a$. $\qquad\square$

Next we define a generalization of the notion of a strong I-algebra:

Definition 2.5. Let $\mathcal{A} = (A, \cdot, 1)$ be an algebra of type $(2, 0)$. $\mathcal{A}$ is called a *weak I-algebra* if it satisfies

$$1x = x, \tag{W1}$$
$$xx = 1, \tag{W2}$$
$$x(yx) = 1, \tag{W3}$$
$$(xy)y = (yx)x \tag{W4}$$

and

$$((xy)y)z = 1 \quad implies \quad xz = 1. \tag{W5}$$

Remark 2.6. (S5) and (S1) imply (W5).

Theorem 2.7. *Within the definition of a weak I-algebra* (W3) *and* (W5) *may be replaced by the laws*

$$((xy)y)y = xy \tag{W3'}$$

and

$$x((((xy)y)z)z) = 1, \tag{W5'}$$

respectively, and hence weak I-algebras form a variety.

Proof. (W4), (W3) and (W1) imply (W3'):

$$((xy)y)y = (y(xy))(xy) = 1(xy) = xy.$$

(W2) and (W5) imply (W5'):

$$(((((xy)y)z)z)((((xy)y)z)z) = 1$$

and hence

$$((xy)y)((((xy)y)z)z) = 1$$

whence

$$x((((xy)y)z)z) = 1.$$

(W3'), (W4) and (W2) imply (W3):

$$x(yx) = ((x(yx))(yx))(yx) = (((yx)x)x)(yx) = (yx)(yx) = 1.$$

Finally, (W1) and (W5') imply (W5):

$$((xy)y)z = 1 \quad \text{implies} \quad xz = x(1z) = x((((xy)y)z)z) = 1. \quad \square$$

Remark 2.8. The variety of weak I-algebras was characterized by the axioms (W1), (W2), (W3'), (W4) and (W5') in [3] where weak I-algebras were called d-implication algebras.

For the next theorem we need some definitions.

Definition 2.9. An algebra $(A, \sqcup)$ of type (2) is called a *directoid* (cf. [10]) if there exists a partial order relation $\leq$ on A such that for all $a, b \in A$, $a \sqcup b$ is an upper bound of a and b that coincides with $\max(a, b)$ if a and b are comparable. $\leq$ is uniquely determined by $\sqcup$ by $x \leq y$ if and only if $x \sqcup y = y$ $(x, y \in A)$. $(A, \sqcup)$ is called *commutative* if $\sqcup$ is commutative. Let $(P, \leq)$ be a poset with smallest element 0 and greatest element 1 and $f: P \to P$. f is called a *switching involution* of $(P, \leq)$ if $f(0) = 1, f(1) = 0$ and $f(f(x)) = x$ or all $x \in P$.

Now we can prove

Theorem 2.10. *Let $\mathcal{A} = (A, \cdot, 1)$ be a weak I-algebra. Define*

$$x \sqcup y := (xy)y \quad \text{and} \quad x^y := xy$$

for all $x, y \in A$. Then $\mathbf{S}(\mathcal{A}) := (A, \sqcup, (^x; x \in A), 1)$ is an algebra such that $(A, \sqcup, 1)$ is a commutative directoid with greatest element 1 and for every $x \in A$ $([x, 1], \leq, {}^x)$ is a poset with a switching involution where $\leq$ denotes the partial order induced by $\sqcup$. Conversely, let $\mathcal{S} := (S, \sqcup, (^x; x \in S), 1)$ be an algebra such that $(S, \sqcup, 1)$ is a commutative directoid with greatest element 1 and for every $x \in S$ $([x, 1], \leq, {}^x)$ is a poset with a switching involution with respect to the partial order induced by $\sqcup$. Define

$$xy := (x \sqcup y)^y$$

for all $x, y \in S$. Then $\mathbf{A}(\mathcal{S}) := (S, \cdot, 1)$ is a weak I-algebra. Moreover, $\mathbf{A}(\mathbf{S}(\mathcal{A})) = \mathcal{A}$ and $\mathbf{S}(\mathbf{A}(\mathcal{S})) = \mathcal{S}$ for every weak I-algebra $\mathcal{A}$ and every algebra $\mathcal{S} = (S, \sqcup, (^x; x \in S), 1)$ such that $(S, \sqcup, 1)$ is a commutative directoid with greatest element 1 and for every $x \in S$ $([x, 1], \leq, {}^x)$ is a poset with a switching involution.

Proof. First assume $\mathcal{A} = (A, \cdot, 1)$ to be a weak I-algebra and for all $x, y \in A$ define $x \leq y$ if $xy = 1$, $x \sqcup y := (xy)y$ and $x^y := xy$. Reflexivity and antisymmetry of $\leq$ follow as in the proof of Theorem 2.4. If $a \leq b \leq c$ then $((ab)b)c = (1b)c = bc = 1$ according to (W1) whence $ac = 1$ by (W5), i.e., $a \leq c$. This shows transitivity of $\leq$. Hence $(A, \leq)$ is a poset. That 1 is the greatest element of $(A, \leq)$ and $\sqcup$ is commutative follows as in the proof of Theorem 2.4. Because of (W2) we have $((ab)b)((ab)b) = 1$ whence by (W5) it follows $a((ab)b) = 1$, i.e., $a \leq a \sqcup b$. Hence $b \leq b \sqcup a = a \sqcup b$. If $a \leq b$ then $a \sqcup b = (ab)b = 1b = b$ according to (W1). This shows that $(A, \sqcup, 1)$ is a commutative directoid with 1. That for all $a \in A$, ^{a} is an involution of $[a, 1]$ follows as in the proof of Theorem 2.4. Finally, $a^a = aa = 1$ according to (W2) and $1^a = 1a = a$ according to (W1) showing that ^{a} is switching. Therefore $\mathbf{S}(\mathcal{A}) = (A, \sqcup, (^a; a \in A), 1)$ is an algebra such that $(S, \sqcup, 1)$ is a commutative directoid with greatest element 1 and for every $x \in S$ $([x, 1], \leq, ^x)$ is a poset with a switching involution. Moreover, $((ab)b)b = ab$ follows as in the proof of Theorem 2.4 showing that $\mathbf{A}(\mathbf{S}(\mathcal{A})) = \mathcal{A}$.

Conversely, assume $\mathcal{S} = (A, \sqcup, (^a; a \in A), 1)$ to be an algebra such that $(S, \sqcup, 1)$ is a commutative directoid with greatest element 1 and for every $x \in S$ $([x, 1], \leq, ^x)$ is a poset with a switching involution. Moreover, for all $x, y \in A$ define $xy := (x \sqcup y)^y$.

(W1)–(W4) follow as in the proof of Theorem 2.4.

(W5) If $((ab)b)c = 1$ then

$$a \leq a \sqcup b \leq (a \sqcup b) \sqcup c = ((((a \sqcup b)^b)^b \sqcup c)^c)^c$$

$$= ((((a \sqcup b)^b \sqcup b)^b \sqcup c)^c)^c = (((ab)b)c)^c = 1^c = c,$$

which implies $ac = (a \sqcup c)^c = c^c = 1$. Therefore $\mathbf{A}(\mathcal{S}) = (A, \cdot, 1)$ is a weak I-algebra. $((a \sqcup b)^b \sqcup b)^b = a \sqcup b$ follows as in the proof of Theorem 2.4 and, moreover, $a \leq b$ implies $(b \sqcup a)^a = b^a$ showing that $\mathbf{S}(\mathbf{A}(\mathcal{S})) = \mathcal{S}$. $\qquad\square$

Remark 2.11. In [3] commutative directoids with greatest element 1 such that for every $x \in S$ $([x, 1], \leq, ^x)$ is a poset with a switching involution were called *commutative directoids with sectional antitone involutions*.

3. Congruence Kernels

The aim of this section it to characterize congruence kernels of weak I-algebras having certain additional properties. First we observe that weak I-algebras are *weakly regular* which means that congruences are determined by the class of 1:

Lemma 3.1. *Let* $\mathcal{A} = (A, \cdot, 1)$ *be a weak I-algebra and* $\Theta \in \mathrm{Con}\,\mathcal{A}$. *Then* $\Theta = \{(x, y) \in A^2 \mid xy, yx \in [1]\Theta\}$.

Proof. If for $a, b \in A$ it holds $a \ominus b$ then $ab \ominus aa = 1$ and $ba \ominus aa = 1$ and if, conversely, $ab \ominus 1$ and $ba \ominus 1$ then $a = 1a \ominus (ba)a = (ab)b \ominus 1b = b$. □

Next we define the notion of a congruence kernel of a weak *I*-algebra:

Definition 3.2. A subset K of the base set A of a weak *I*-algebra $\mathcal{A}$ is called a *congruence kernel* of $\mathcal{A}$ if there exists a congruence $\Theta \in \mathrm{Con}\,\mathcal{A}$ with $[1]\Theta = K$. Let $\mathrm{Ker}\,\mathcal{A}$ denote the set of all congruence kernels of $\mathcal{A}$.

Theorem 3.3. *The mappings* $\Theta \mapsto [1]\Theta$ *and* $K \mapsto \{(x,y) \in A^2 \mid xy, yx \in K\}$ *are mutually inverse isomorphisms between* $(\mathrm{Con}\,\mathcal{A}, \subseteq)$ *and* $(\mathrm{Ker}\,\mathcal{A}, \subseteq)$ *and hence the latter is a complete lattice.*

Proof. The proof follows immediately from Lemma 3.1. □

Certain subsets of weak *I*-algebras have a nice property which will be used in the proof of the final theorem of this section. In the following for a subset K of a weak *I*-algebra $(A, \cdot, 1)$ and an element a of A define $Ka := \{ka \mid k \in K\}$. More generally, for subsets K, L of A we define $KL := \{kl \mid k \in K, l \in L\}$.

Lemma 3.4. *Let* $\mathcal{A} = (A, \cdot, 1)$ *be a weak I-algebra, K a subset of A, $a \in K$ and $b \in A$ and assume $ab \in K$ and $(K(Kx))x \subseteq K$ for all $x \in A$. Then $b \in K$.*

Proof. $b = 1b = (a((ba)a))b = (a((ab)b))b \in (K(Kb))b \subseteq K$. □

Now we can state and prove the characterization of congruence kernels of certain weak *I*-algebras:

Theorem 3.5. *Let* $\mathcal{A} = (A, \cdot, 1)$ *be a weak I-algebra satisfying*

$$x(yz) = (xy)(xz) \qquad and \qquad (xy)((yz)(xz)) = 1$$

and let K be a subset of A. Then K is a congruence kernel of $\mathcal{A}$ if and only if $1 \in K$, $AK \subseteq K$ and $(K(Kx))x \subseteq K$ for all $x \in A$.

Proof. Let $a, b, c \in A$ and $d, e \in K$.

First assume $K \in \mathrm{Ker}\,\mathcal{A}$. Then there exists a $\Theta \in \mathrm{Con}\,\mathcal{A}$ with $[1]\Theta = K$ and hence

$$1 \in [1]\Theta = K,$$
$$ad \in [a1]\Theta = [a(aa)]\Theta = [1]\Theta = K$$

and

$$(d(ea))a \in [(1(1a))a]\Theta = [(1a)a]\Theta = [aa]\Theta = [1]\Theta = K$$

proving $1 \in K$, $AK \subseteq K$ and $(K(Kx))x \subseteq K$ for all $x \in A$.

Conversely, assume $1 \in K$, $AK \subseteq K$ and $(K(Kx))x \subseteq K$ for all $x \in A$. Put $\Phi := \{(x, y) \in A^2 \mid xy, yx \in K\}$.

Since $1 \in K$, Φ is reflexive.

Obviously, Φ is symmetric.

If $a \ \Phi \ b \ \Phi \ c$ then $ab, ba, bc, cb \in K$ and hence $a(bc) \in AK \subseteq K$ and $(a(bc))(ac) = ((ab)(ac))(ac) = (1((ab)(ac)))(ac) \in (K(K(ac)))(ac) \subseteq K$ whence $ac \in K$ according to Lemma 3.4. Interchanging the roles of a and c yields $ca \in K$ and hence $a \ \Phi \ c$. This shows transitivity of Φ.

If $a \ \Phi \ b$ then $ab, ba \in K$ and hence $(ab)((bc)(ac)) = 1 \in K$ whence $(bc)(ac) \in K$ according to Lemma 3.4. Interchanging the roles of a and b yields $(ac)(bc) \in K$ and hence $ac \ \Phi \ bc$. This shows that Φ is a right congruence on $\mathcal{A}$.

If, finally, $a \ \Phi \ b$ then $ab, ba \in K$ and hence $(ca)(cb) = c(ab) \in AK \subseteq K$. Interchanging the roles of a and b yields $(cb)(ca) \in K$ and hence $ca \ \Phi \ cb$. This shows that Φ is a left congruence on $\mathcal{A}$.

Altogether we have proved $\Phi \in \mathrm{Con}\,\mathcal{A}$. Since, obviously, $[1]\Phi = K$, the proof of the theorem is complete. $\square$

4. Varieties of Implication Algebras

In this section we prove that the different varieties of implication algebras mentioned within the paper form a strictly increasing chain with respect to inclusion.

Definition 4.1. Let V_1, V_2, V_3, V_4 and V_5 denote the variety of all implication algebras, orthoimplication algebras, orthomodular implication algebras, strong I-algebras and weak I-algebras, respectively.

Theorem 4.2. $V_1 \subset V_2 \subset V_3 \subset V_4 \subset V_5$.

Proof. The orthoimplication algebra corresponding to $\mathcal{MO}_2 := 2^2 + 2^2$ belongs to $V_2 \setminus V_1$ since $[0, 1]$ is not a Boolean algebra. (Here $+$ denotes the horizontal sum.) The orthomodular implication algebra corresponding to $\mathcal{MO}_2 \times 2^1$ with the Hasse diagram

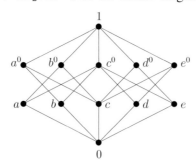

and with $(a^0)^c := d^0$ and $(b^0)^c := e^0$ belongs to $V_3 \setminus V_2$ since $(a^0)^c = d^0 \neq b^0 = a \vee c = (a^0)^0 \vee c$. The strong I-algebra corresponding to the three-element chain belongs to $V_4 \setminus V_3$ since $[0, 1]$ is not an orthomodular lattice. The weak I-algebra corresponding to the poset with the Hasse diagram

with $a \sqcup b := 1$ belongs to $V_5 \setminus V_4$ since it is not a join-semilattice. $\square$

References

[1] ABBOTT, J. C. (1967) Semi-boolean algebra. Mat. Vesnik **4**: 177–198

[2] ABBOTT, J. C. (1976) Orthoimplication algebras. Studia Logica **35**: 173–177

[3] CHAJDA, I. (2007) Commutative directoids with sectional involutions. Discussiones Math. **27**: 49–58

[4] CHAJDA, I. (2007) Ring-like structures derived from λ-lattices with antitone involutions. Math. Bohemica **132**: 87–96

[5] CHAJDA, I., EIGENTHALER, G. (2007) Semilattices with sectional mappings. Discussiones Math. **27**: 11–19

[6] CHAJDA, I., EMANOVSKÝ, P. (2004) Bounded lattices with antitone involutions and properties of MV-algebras. Discussiones Math. **24**: 31–42

[7] CHAJDA, I., HALAŠ, R., KÜHR, J. (2005) Distributive lattices with sectionally antitone involutions. Acta Sci. Math. (Szeged) **71**: 19–33

[8] CHAJDA, I., HALAŠ, R., LÄNGER, H. (2001) Orthomodular implication algebras. Intern. J. Theor. Phys. **40**: 1875–1884

[9] CHAJDA, I., HALAŠ, R., LÄNGER, H. (2004) Simple axioms for orthomodular implication algebras. Intern. J. Theor. Phys. **43**: 911–914

[10] JEŽEK, J., QUACKENBUSH, R. (1990) Directoids: Algebraic models of up-directed sets. Algebra Universalis **27**: 49–69

Authors' addresses: Ivan Chajda, Department of Algebra and Geometry, Palacký University Olomouc, Tomkova 40, 77900 Olomouc, Czech Republic. E-Mail: chajda@inf.upol.cz; Helmut Länger, Institute of Discrete Mathematics and Geometry, Vienna University of Technology, Wiedner Hauptstraße 8–10, 1040 Vienna, Austria. E-Mail: h.laenger@tuwien.ac.at.

Sitzungsber. Abt. II (2007) 216: 45–52

Sitzungsberichte

Mathematisch-naturwissenschaftliche Klasse Abt. II
Mathematische, Physikalische und Technische Wissenschaften

Reliable Estimates of Planetary and Solar Magnetic Fields – A Case for Abundance of Hydrogen in the Earth's Core

By

Fritz Paschke

(Vorgelegt in der Sitzung der math.-nat. Klasse am 15. November 2007
durch das w. M. Fritz Paschke)

Abstract

A theory of planetary magnetic fields published in 1998, which avoids the drawbacks of the conventional dynamo theories and is based on rotating dipole domains, is supplemented by adding data for the Sun and young Mars. Acceptable accuracy is achieved in the prediction of magnetic moments spanning 11 orders of magnitude for celestial bodies differing in mass by a factor of $6 \cdot 10^6$. One of the weaknesses of the theory lies in the fact that in assuming iron to be the dominant element in the Earth's core, inacceptable temperature levels are necessary to support the dipole model. However, on the basis of later reports it may be concluded that metallic hydrogen is present in the Earth's core in a higher concentration than assumed earlier. Taking the presence of hydrogen into account, the temperatures needed to support the dipole model are reduced to widely accepted levels, provided that the ratio of hydrogen on the total mass density is at least $1/6$ (calculated for a valency of the iron nuclei of $Z = 1$) or at least $1/10$ (estimated for $Z = 2$). Thus Earth-like planets appear closer in structure to the gas planets than assumed hitherto. A loss of hydrogen in the core leads to the extinction of the magnetic field – a fate, which may have been suffered by Mars.

1. Introduction

The model used in ref. [1] is based on considering a slab of conducting liquid keeping together by a binding energy represented in Fig. 1 by a potential wall.

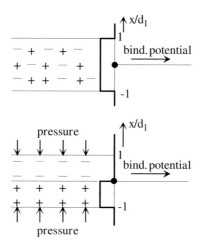

Fig. 1. *Upper part:* Conducting plasma layer of thickness $2d_1$, binding energy symbolized by potential barriers. *Lower part*: Plasma layer under high pressure, forming dipole domains. Here d_1 is 3,3 times the Debye-distance (Eq. (19) in ref. [1]), calculated for the average particle density

The assumption made to obtain a loss-free model is that under high pressure the electrons separate from the densely packed atomic nuclei and form a dipole layer. The large force due to electric space-charge fields is balanced by pressure gradients as indicated for a valency of $Z = 1$ in Figs. 2 and 3.

To prevent back-diffusion of the electrons, the potential barrier has to be maintained at a level $\gg kT/e$, with k as Boltzmann's constant,

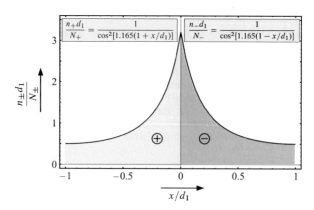

Fig. 2. Distribution of particle densities within the dipole domains relative to the average particle density (Eqs. (18) and (10) of ref. [1]), with $N_\pm$ as particle density per area

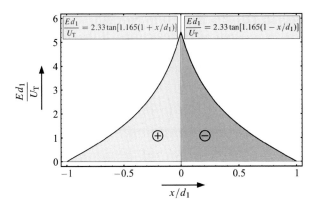

Fig. 3. Distribution of the electric field within the dipole domains (Eqs. (7) and (9) of ref. [1]), normalized to U_T/d_1

T as absolute temperature and e as elementary electric charge. Note that the dipole layer maintains neutrality towards the outside (zero external electric field).

When stacking up such dipole layers onion-like for the planetary cores and attributing to the particles an additional degree of freedom based on the stored electrical energy per particle, the following formulas for the magnetic moments m_H had been derived in ref. [1].

For a spherical shell of thickness da at the radius a (Eq. (28) in ref. [1])

$$\frac{dm_H}{2\pi} = -1.5\Omega \cdot \sqrt{p_0\varepsilon} \cdot a^3 da. \tag{1}$$

Here ε is the permittivity, Ω is the angular velocity (for Earth $\Omega > 0$) and p_0 is the pressure, which is approximated by the parabolic function (Eq. (29) in ref. [1])

$$p_0(a) = p_0(0) - \left(\frac{a}{a_1}\right)^2 (p_0(0) - p_0(a_1)) \tag{2}$$

with a_1 the outer radius of the core. Integrating Eq. (1) with Eq. (2) over the core yields the total magnetic dipole moment (Eq. (31) in ref. [1])

$$\frac{m_H}{2\pi} = -\frac{8\Omega}{3}\left(\frac{\varepsilon p_0(0)}{2}\right)^{1/2} \cdot 0.8a_1^4$$

$$\times \frac{\frac{1}{3}\left[1 - \left(\frac{p_0(a_1)}{p_0(0)}\right)^{3/2}\right] - \frac{1}{5}\left[1 - \left(\frac{p_0(a_1)}{p_0(0)}\right)^{5/2}\right]}{\left(1 - \frac{p_0(a_1)}{p_0(0)}\right)^2}. \tag{3}$$

For the signs in Eqs. (1) and (3) to be valid, it is assumed that the outermost layer is composed of electrons. If the dipole orientation is reversed (ions in the outermost layer) the field is reversed.

Eqs. (1) and (2) allow the calculation of the magnetic fields. With

$$H_0 = -\frac{8}{3} \cdot \Omega \cdot \left(\frac{\varepsilon p_0(0)}{2} \right)^{1/2} \cdot 0.8 a_1 \tag{4}$$

and the normalized radius

$$\rho = \frac{r}{a_1} \tag{5}$$

the distributions in spherical coordinates for $\rho \leq 1$ read

$$H_r = H_0 \cdot \cos\vartheta \left\{ \frac{1}{k^2\rho^3} \left[\frac{1}{3}[1-(1-\rho^2 k)^{3/2}] - \frac{1}{5}[1-(1-\rho^2 k)^{5/2}] \right] \right.$$
$$\left. + \frac{1}{2k} \left[\arcsin k - \arcsin\rho k + k(1-k)^{1/2} - \rho k(1-\rho^2 k)^{1/2} \right] \right\}, \tag{6a}$$

$$H_\vartheta = \frac{1}{2} \cdot H_0 \cdot \sin\vartheta \cdot \left\{ \frac{1}{k^2\rho^3} \left[\frac{1}{3}[1-(1-\rho^2 k)^{3/2}] - \frac{1}{5}[1-(1-\rho^2 k)^{5/2}] \right] \right.$$
$$\left. - \frac{1}{k} \left[\arcsin k - \arcsin\rho k + k(1-k)^{1/2} - \rho k(1-\rho^2 k)^{1/2} \right] \right\} \tag{6b}$$

and for $\rho \geq 1$

$$H_r = H_0 \cdot \cos\vartheta \cdot \frac{1}{\rho^3 k^2} \left[\frac{1}{3}[1-(1-k)^{3/2}] - \frac{1}{5}[1-(1-k)^{5/2}] \right], \tag{7a}$$

$$H_\vartheta = H_0 \cdot \sin\vartheta \cdot \frac{1}{2\rho^3 k^2} \left[\frac{1}{3}[1-(1-k)^{3/2}] - \frac{1}{5}[1-(1-k)^{5/2}] \right]. \tag{7b}$$

Here, not to be confounded with Boltzmann's constant introduced later,

$$k = 1 - \frac{p_0(a_1)}{p_0(0)}. \tag{8}$$

For Earth, $k = 0.6$ and the field distribution of Fig. 4 is obtained, with the magnetic south pole at the geographic north pole. The energy of the Earth magnetic field is calculated to be $4.563 \cdot 10^{18}$ Joule, 14.3% of which is stored outside of the core.

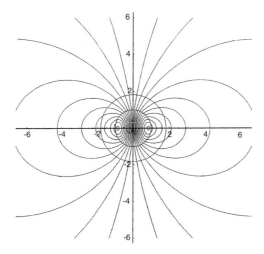

Fig. 4. Field lines, calculated from Eqs. (6) and (7) for Earth, with the radii relative to the outer core radius

2. Magnetic Moments in the Solar System

Eq. (7a) leads to an Earth magnetic field at the poles of about 24 A/m, which corresponds to an induction of 30 μT (0.3 Gauss); this is close to the observed values. When applying the model to other planets, the lack of knowledge of their internal structures is a problem which had been tried to overcome by using a two-shell model with a core material of average mass-density ρ_1 and a mantle material of average mass-density ρ_2. With plausible values for $\rho_{1,2}$ and the known values of radii and masses, the parameters entering Eq. (3) can be determined. Confidence in this method is gained by applying it to Earth, where the two-shell model leads to an error of only 2% in magnetic moment as compared to the result gained from the more accurate data available from seismic measurements.

The values of the mass densities used were:

for Earth and Venus $\rho_1 = 1.2 \cdot 10^4 \, \text{kg/m}^3$, $\rho_2 = 4.22 \cdot 10^3 \, \text{kg/m}^3$;

for Mercury and Mars $\rho_1 = 10^4 \, \text{kg/m}^3$, $\rho_2 = 3 \cdot 10^3 \, \text{kg/m}^3$;

for Jupiter and Saturn $\rho_1 = 1.4 \cdot 10^4 \, \text{kg/m}^3$, $\rho_2 = 0.55 \cdot 10^3 \, \text{kg/m}^3$;

for Uranus and Neptune $\rho_1 = 1.3 \cdot 10^4 \, \text{kg/m}^3$, $\rho_2 = 0.8 \cdot 10^3 \, \text{kg/m}^3$,

for Sun $\rho_1 = 1.3 \cdot 10^5 \, \text{kg/m}^3$, $\rho_2 \ll \rho_1$.

Table 1. $a_1/10^6$ m is the core radius in 10^6 Meters, $p(a_1)/10^{11}$ P and $p(0)/10^{11}$ P are the pressures at the edge and the center of the core in 10^{11} Pascal, Ω/Ω_E the angular velocity of the planet related to the Earth's value, and m_H/m_{HE} the magnetic dipole moment related to the Earth's value

	$\dfrac{a_1}{10^6 \text{ m}}$	$\dfrac{p(a_1)}{10^{11}\text{ P}}$	$\dfrac{p_0}{10^{11}\text{ P}}$	$\dfrac{\Omega}{\Omega_E}$	$\dfrac{m_H}{m_{HE}}$ theor.	$\dfrac{m_H}{m_{HE}}$ observed, from NESS [2]
Mercury	1.71	0.0633	0.513	$1.7 \cdot 10^2$	$2 \cdot 10^{-4}$	$6.25 \cdot 10^{-4}$
Venus	3.0	1.09	3.14	$4.1 \cdot 10^{-3}$	$2.5 \cdot 10^{-3}?/0$	$<5 \cdot 10^{-5}$
Earth	3.5	1.21	4.6	1	1	1
Mars	1.7	0.2	0.6	0.975	$2 \cdot 10^{-2}$	$(<2.5 \cdot 10^{-4}$ today) $2 \cdot 10^{-2}$ from Mars Global Surveyor for young Mars
Jupiter	26.5	8.61	200	2.41	$4.2 \cdot 10^4$	$2 \cdot 10^4$
Saturn	10.7	3.41	34.6	2.24	463	590
Uranus	8.28	1.8	17.9	1.39	74.3	47.5
Neptune	9.82	2.07	24.7	1.48	180	25
Sun	152.4	$\ll p(0)$	$5.48 \cdot 10^5$	$4 \cdot 10^{-2}$	$2.18 \cdot 10^7$	10^7 (from ref. [3])

Table 1 summarizes the results for the solar system (excluding Pluto). No attempt has been made to improve agreement by adopting other values for $\rho_{1,2}$ than those given above. The agreement of observed (N. F. NESS [2]) and calculated values spanning 11 orders of magnitude for celestial bodies differing in mass by $6 \cdot 10^6$ is remarkable. The data given for the Sun refer to the dipole-part of its magnetic field (FRIEDMANN [3], MERRILL et al. [4]). Also remarkable appears the fact that, based on observations of the Global Surveyor (CONNERNEY et al. [5]), young Mars had exactly the magnetic moment predicted by the theory, but lost it later for unknown reasons. For Venus, the rotational speed $(<0.9\,\text{m/s})$ is well below the thermal velocities, so that the magnetic moment can be assumed to be zero – Eq. (3) is not applicable.

3. The Temperature Dilemma

The theory suffers from the drawback that very high core temperatures are necessary to support the dipole model. For a valency of $Z = 1$ and predominating iron, the temperature (from Eq. (34) in ref. [1]) reads

$$T = \frac{m \cdot p_0}{\rho_m \cdot k} \cdot \frac{2}{1+Z} \tag{9}$$

with ρ_m as mass density, m as atomic mass (iron for Earth), p_0 as pressure, Z as valency and k as Boltzmann's constant. With $m = 9.37 \cdot 10^{-26}$ kg, $\rho_m \doteq 10^4$ kg/m^3, $p_0 = 1.6 \cdot 10^{11}$ P – the value at the outer edge of the Earth's core – and a valency of $Z = 1$, an unreasonable value of $T = 109{,}020$ Kelvin is obtained. A great dilemma for the theorist, who may be rescued by well-based arguments (WILLIAMS and HEMLEY [6]) for an abundance of hydrogen in the core.

Thus Eq. (9) has been re-examined for a mixture of iron and hydrogen. The field- and charge distributions, calculated in ref. [1], are valid for a mixture only for equal valency of the components. Thus a valency of $Z = 1$ had to be assumed for the iron nuclei. Taking $0 \le \gamma \le 1$ as fraction of the hydrogen part on the total mass density,

$$\rho_m = \frac{mN_+}{d_1} \cdot \frac{1}{1 + \gamma\left(\dfrac{m}{m_p} - 1\right)} \tag{10}$$

with m_p as the proton mass and N_+/d_1 as average density of the nuclei, which, from Eqs. (12) and (15) of ref. [1], is given by

$$\frac{N_+}{d_1} = \frac{2p_0}{kT}. \tag{11}$$

From Eqs. (10) and (11) the temperature becomes

$$T = \frac{mp_0}{\rho_m \cdot k} \cdot \frac{1}{1 + \gamma\left(\dfrac{m}{m_p} - 1\right)}. \tag{12}$$

The reduction factor for the temperature reaches the necessary value of (at least) 0.1 for $\gamma = 0.16$ or $1/6$. If a reduction down to 5000 Kelvin were required, $\gamma \doteq 0.38$. For valencies of the iron nuclei exceeding 1, the necessary hydrogen concentration becomes lower, but cannot be analyzed rigorously on the basis of ref. [1]. But a crude estimate is given by Eq. (9), which yields a reduction factor for the temperature of $2/(1 + Z)$, so that with $Z = 2, \gamma = 0.1$ or $1/10$ is estimated for a reduction factor of 0.1.

The high concentration of hydrogen postulated here appears possible in a plasma-state of the material. Metallic hydrogen caught the interest of scientists since a long time to explain, for example, the behaviour of gas planets or to seek practical applications (W. J. NESS [7]). For the pressure inside of the Earth's core, a plasma state appears possible. Critical is the application for Mercury, where the pressure reaches only about 50 GP (see Table 1), but metallic hydrogen is rich in unexpected properties (BABAEV et al. [8]), so that a mixture of iron and hydrogen may be suspected to be in a plasma state even at the pressures in Mercury's core.

The conclusions are:

- The theory of rotating dipole domains yields reliable estimates of the magnetic moments for Sun and all planets.
- For reasonable core temperatures, hydrogen has to appear in abundance in all Earth-like planets.
- For Mars, the loss of magnetic field may have been caused by a loss of hydrogen in its core.

References

[1] PASCHKE, F. (1998) Rotating electric dipole domains as a loss-free model for the earth's magnetic field. Sitzungsber. Abt. II **207**: 213–228; Transact. Austrian Academy of Sciences

[2] NESS, N. F. (1994) Intrinsic magnetic fields of the planets: Mercury to Neptune. Phil. Trans. R. Soc. London. Ser. A **349**: 249–260

[3] FRIEDMAN, H. (1986) Sun and Earth. Sci. Americ. Books, New York

[4] MERRILL, R. T., MCELHINNY, M. W., MCFADDEN, P. L. (1998) The Magnetic Field of the Earth (Int. Geophys. Ser., Vol. 63). Academic Press, San Diego

[5] CONNERNEY, J. E. P., ACUNA, M. H., WASILEWSKI, P. J., NESS, N. F., RÈME, H., MAZELLE, C., VIGNES, D., LIN, R. P., MITCHELL, D. L., CLOUTIER, P. A. (1999) Magnetic lineations in the ancient crust of mars. Science **284**: 794–798

[6] WILLIAMS, Q., HEMLEY, R. J. (2001) Hydrogen in the deep earth. Ann. Rev. Earth Planet Sci. **29**: 365–418

[7] NESS, W. J. (1999) Metastable solid metallic hydrogen. Phil. Mag. B **79**/4: 655–661

[8] BABAEV, E., SUDBO, A., ASHCROFT, N. W. (2004) A superconductor to superfluid phase transition in liquid metallic hydrogen. Nature **431**: 666–668

Author's address: Prof. Dr. Fritz Paschke, Kahlenberger Straße 35/2, 1190 Wien, Austria. E-Mail: fritz.paschke@chello.at.

Sitzungsber. Abt. II (2007) 216: 53–56

Sitzungsberichte

Mathematisch-naturwissenschaftliche Klasse Abt. II
Mathematische, Physikalische und Technische Wissenschaften

© Österreichische Akademie der Wissenschaften 2008
Printed in Austria

The Invariant Measure
for the Two-Dimensional Parry-Daniels Map

By

Fritz Schweiger

(Vorgelegt in der Sitzung der math.-nat. Klasse am 15. November 2007
durch das k. M. I. Fritz Schweiger)

Abstract

The Parry-Daniels map T has an exceptional set Γ which can be seen as a strange attractor for T. The density of the invariant measure is given. Some remarks on the exceptional set for the mixture of the Selmer algorithm and the fully subtractive algorithm are added.

Mathematics Subject Classification (2000): 11K55, 28D99.
Key words: Ergodic theory, invariant measures.

Let $x = (x_0, x_1, x_2) \in (\mathbb{R}^+)^3$ and let π be a permutation of the indices such that $x_{\pi 0} \leq x_{\pi 1} \leq x_{\pi 2}$. The Poincaré map P is defined as

$$P(x_0, x_1, x_2) = (x_{\pi 0}, x_{\pi 1} - x_{\pi 0}, x_{\pi 2} - x_{\pi 1}).$$

We introduce

$$\Sigma^2 = \{x \in (\mathbb{R}^+)^3 : x_0 + x_1 + x_2 = 1\}.$$

Then the Parry-Daniels map $T \colon \Sigma^2 \to \Sigma^2$ is defined as

$$T(x_0, x_1, x_2) = \left(\frac{x_{\pi 0}}{x_{\pi 2}}, \frac{x_{\pi 1} - x_{\pi 0}}{x_{\pi 2}}, \frac{x_{\pi 2} - x_{\pi 1}}{x_{\pi 2}} \right),$$
$$\pi \in \{\varepsilon, (01), (02), (12), (012), (021)\}.$$

We introduce the notation

$$x^{(k)} = \left(x_0^{(k)}, x_1^{(k)}, x_2^{(k)}\right) := P^k x.$$

We define

$$\sigma(x) := \sum_{k \geq 0} \max\left(x_0^{(k)}, x_1^{(k)}\right).$$

The following result could be proved (SCHWEIGER [2], NOGUEIRA [1]).
Let

$$\Gamma := \bigcap_{s=0}^{\infty} \bigcup_{\pi_1,\ldots,\pi_s \in \{\varepsilon,(01)\}} B(\pi_1, \ldots, \pi_s),$$

then $\Gamma = \{x \in \Sigma^2 : \sigma(x) \leq x_2\}$ and $\lambda(\Gamma) > 0$. Since T is ergodic with respect to Lebesgue measure, we obtain

$$\Sigma^2 = \bigcup_{j=0}^{\infty} T^{-j} \Gamma.$$

Note that $\sigma(x)$ is convergent for almost all directions $\theta = x_0/x_1$ or $\theta = x_1/x_0$, $0 \leq \theta \leq 1$.

Since on Σ^2 the relation $x_2 = 1 - x_0 - x_1$ holds, we restrict our attention to the first coordinates, i.e. to the domain $\{(x_0, x_1) : 0 \leq x_0, 0 \leq x_1, 0 \leq x_0 + x_1 \leq 1\}$.

Theorem. *The function*

$$h(x_0, x_1) = \frac{1}{x_0(x_0 + x_1)(1 - x_0 - x_1 - \sigma(x_0, x_1))}$$

is an invariant density for T restricted to Γ.

Proof. The map T restricted to Γ has only two inverse branches

$$V(\varepsilon)(x_0, x_1) = \left(\frac{x_0}{1 + 2x_0 + x_1}, \frac{x_0 + x_1}{1 + 2x_0 + x_1}\right),$$

$$V(01)(x_0, x_1) = \left(\frac{x_0 + x_1}{1 + 2x_0 + x_1}, \frac{x_0}{1 + 2x_0 + x_1}\right).$$

Then

$$h(V_0(\varepsilon)(x_0, x_1))\omega(\varepsilon, x_0, x_1) + h(V(01)(x_0, x_1))\omega(01; x_0, x_1)$$

$$= \frac{1}{x_0(2x_0 + x_1)\left(1 - (1 + 2x_0 + x_1)\sigma\left(\dfrac{x_0}{1 + 2x_0 + x_1}, \dfrac{x_0 + x_1}{1 + 2x_0 + x_1}\right)\right)}$$

$$+\frac{1}{x_0(2x_0+x_1)\left(1-(1+2x_0+x_1)\sigma\left(\dfrac{x_0}{1+2x_0+x_1},\dfrac{x_0+x_1}{1+2x_0+x_1}\right)\right)}.$$

We note the following properties of the function σ:

$$\sigma(\lambda y_0,\lambda y_1) = \lambda\sigma(y_0,y_1),$$
$$\sigma(y_0,y_1) = \sigma(y_1,y_0),$$
$$\sigma(x_0,x_0+x_1) = x_0+x_1+\sigma(x_0,x_1).$$

Therefore

$$(1+2x_0+x_1)\sigma\left(\frac{x_0}{1+2x_0+x_1},\frac{x_0+x_1}{1+2x_0+x_1}\right) = x_0+x_1+\sigma(x_0,x_1).$$

Hence

$$h(V(\varepsilon)(x_0,x_1))\omega(\varepsilon;x_0,x_1)+h(V(01)(X_0,x_1))\omega(01;x_0,x_1)=h(x_0,x_1).$$

Remark 1. The set Γ can be described as consisting of all needles emanating from $(0,0)$ which are given by the equations

$$x_0 = \lambda, \qquad x_1 = \lambda\theta, \qquad 0 \le \lambda \le \frac{1}{1+\theta+S(\theta)}$$

or

$$x_0 = \lambda\theta, \qquad x_1 = \lambda, \qquad 0 \le \lambda \le \frac{1}{1+\theta+S(\theta)},$$

$$S(\theta) = \sigma(\theta,1) = \sigma(1,\theta), \qquad 0 \le \theta \le 1.$$

Therefore the equation

$$x_0+x_1+\sigma(x_0,x_1) = 1$$

can be viewed as referring to the boundary of Γ in some sense (the other parts of the boundary are given by $x_0 = 0$ and $x_1 = 1$).

Remark 2. This remark concerns the paper SCHWEIGER [3]. In this paper the Selmer algorithm S and the Fully Subtractive algorithm T were considered. The following theorem was proved:

Theorem. Let $\Gamma = (x_1,x_2)\in B^2 : (S \circ T)^j x\in E,\ j \ge 0$. Then $\lambda(\Gamma)>0$.

The proof given was a modification of SCHWEIGER [2]. The essential idea is to show that

$$\frac{q_n}{A_n} \ge \gamma>0, \qquad \gamma = \gamma(u) \quad \text{a.e.}$$

However in contrast to the Parry-Daniels map it is easy to show that there is a constant $\gamma > 0$ such that for all u

$$\frac{q_n}{A_n} \geq \gamma > 0.$$

From

$$a_{n+1} \leq \frac{q_{n+1}}{q_n}$$

one sees by induction that

$$q_n \leq \left(2 + \frac{1}{q_1} + \cdots + \frac{1}{q_{n-1}} \right) A_n$$

holds. This implies that the set Γ contains a triangle. Therefore the set Γ is less "exceptional" as explained in Remark 2. In fact, Γ contains the triangle with the vertices $(0,0)$, $\left(\frac{1}{2}, 0 \right)$, $\left(\frac{1}{3}, \frac{1}{3} \right)$. But it is easy to see that Γ contains at least countably many segments which start at $(0,0)$ but go beyond the line $2x_1 + x_2 = 1$.

The restriction of $S \circ T$ on Γ has the σ-finite invariant measure with density

$$h(x_1, x_2) = \frac{1}{x_1 x_2 (1 - 2x_1 - x_2)}.$$

Acknowledgement

This paper was inspired by discussions on the dynamics of T on Γ during the Workshop on Dynamical Systems and Number Theory in Strobl (July 2007).

References

[1] NOGUEIRA, A. (1995) The three-dimensional Poincaré continued fraction algorithm. Israel J. Math. **90**: 373–401

[2] SCHWEIGER, F. (1981) On the Parry-Daniels transformation. Analysis **1**: 171–175

[3] SCHWEIGER, F. (2004) Ergodic and Diophantine properties of algorithms of Selmer type. Acta Arith. **114**: 99–111

Author's address: Prof. Dr. Fritz Schweiger, Department of Mathematics, University of Salzburg, Hellbrunner Strasse 34, 5020 Salzburg, Austria. E-Mail: fritz.schweiger@sbg.ac.at.

Sitzungsber. Abt. II (2007) 216: 57–126

Sitzungsberichte

Mathematisch-naturwissenschaftliche Klasse Abt. II
Mathematische, Physikalische und Technische Wissenschaften

© Österreichische Akademie der Wissenschaften 2008
Printed in Austria

Kleine Planeten, deren Namen einen Österreichbezug aufweisen (II)*

Ein Beitrag zum 160. Jahrestag der Gründung der Akademie der Wissenschaften 2007

Von

Hermann Haupt und Gerhard Hahn

(Vorgelegt in der Sitzung der math.-nat. Klasse am 13. Dezember 2007
durch das w. M. Hermann Haupt)

Abstract

Since twelve years, when paper I "Minor Planets Whose Names Show a Connection to Austria" was published by A. SCHNELL and H. HAUPT (1995), a considerable increase of newly detected bodies has taken place. About 170,000 have been numbered so far, from which more than 14,000 were named, but for only 175 we found a connection to Austria. Now, these latter ones were considered carefully and consequently have been put in the main list of our paper giving all relevant data. The *citation* is given *in English* as published in the official Minor Planet Circulars. Notes, additions and sometimes corrections have then be added in *German* language and will shed some light upon the cultural relations of discoverers and the names of people and places in Austria. It thus turns out, that the members of the Austrian Academy of Sciences (ÖAW) play an important role for the past and now in populating the Minor Planet Belts with their names.

1. Einleitung

Vor zwölf Jahren haben wir zum 1000-jährigen Österreich-Jubiläum und zur Feier des 150. Jahrestages der Gründung der Akademie eine

---

* Herrn Astronomiedirektor Dr. L. D. SCHMADEL zum 65. Geburtstag gewidmet.

Arbeit mit dem gleichen Titel (Teil I: A. SCHNELL und H. HAUPT (1995)) vorgelegt. Darin sind die bis dahin mit einem Österreichbezug benannten Kleinplaneten gelistet und adäquat beschrieben worden. Das Ganze wurde einbegleitet durch eine ausführliche Beschreibung unseres Anliegens und unseres Vorgehens. Das kann hier natürlich nicht in aller Länge wiederholt werden. Es haben sich aber verschiedene Umstände in der Entdeckung der Kleinkörper verändert, und die große Zahl der Neuaufnahmen macht eine leicht gewandelte Vorgangsweise notwendig. Das Wichtigste muss daher trotzdem kurz zusammengefasst werden:

Es darf vorausgesetzt werden, dass die *Geschichte der Kleinplaneten-Entdeckungen* seit 1801 bekannt ist, mindestens so weit, wie sie in unserer oben angeführten Arbeit dargestellt wurde. Während man ursprünglich nur darauf aus war, möglichst viele Planetoiden zu finden und astrometrisch zu vermessen (d. h. ihre Bahnen zu bestimmen), weil man glaubte, dass es sich um Bruchstücke eines ehemaligen großen Planeten handelte, erlebte die physikalische Forschung in der zweiten Hälfte des vorigen Jahrhunderts eine besondere Blütezeit. Durch Beobachtungen von der Erde und vom Weltraum aus konnte eine Reihe von Parametern bestimmt werden, wie Größe und Form dieser Körper, ihre Rotationseigenschaften, die Beschaffenheit ihrer Oberflächen und ihre chemische Zusammensetzung. Dadurch erhielt man bessere Vorstellungen über diese Objekte und – in Zusammenhang mit der Kometenforschung – die Gewissheit, dass es sich um Materie aus der Entstehungszeit unseres Sonnensystems handelte, die sich entsprechend den himmelsmechanischen Gesetzen in bestimmten Regionen um die Sonne angesammelt hatte. Neuerdings, wo es möglich geworden ist, fotografisch und vor allem mit elektronischen Mitteln (CCD's) zu ganz kleinen und lichtschwachen Objekten vorzudringen, steht die Gefahr eines Zusammenstoßes mit der Erde im Blickpunkt des Interesses.

Es ist vor allem diese systematische Suche nach so genannten erdnahen Objekten („Near-Earth Objects" – NEOs), die die Entdeckungsrate von Kleinplaneten (KP) sprunghaft ansteigen ließ. Betrug die Anzahl aller Asteroiden mit gesicherten Bahnen zur Zeit unseres vorigen Artikels (1995) etwas mehr als 6000, liegt diese Zahl heute (November 2007) bei über 170,000. Von diesen nummerierten KP sind weniger als 10 %, z. Z. etwas mehr als 14,000, mit Namen versehen. Dieser enorme Anstieg an zu benennenden Körpern stellt sowohl an die Entdecker, die Vorschläge für geeignete Namen machen sollen, als auch an das Komitee der IAU, das diese Vorschläge zu begutachten und dann schließlich zu genehmigen hat, große Anforderungen. Es stellt sich die Frage, ob man an der bisherigen Praxis, jedem einzelnen KP einen Namen zu geben, auch in Zukunft festhalten kann. Denn die nächste Generation von geplanten und zum Teil schon gestarteten Suchprogrammen wird die Anzahl von entdeckten KP noch einmal vervielfachen – sodass wir in den nächsten 10–20 Jahren von einigen Millionen Objekten ausgehen können.

In der *vorliegenden Arbeit* geht es um die *Namensgebung* oder *„Taufe"* dieser kleinen Planeten. Die bahnmäßig gesicherten Objekte werden nämlich mit endgültigen Nummern versehen, und der Entdecker oder der definitive Bahnrechner hat das Recht, sie zu benennen. Anfangs kamen – wie bekannt – Namen aus der Mythologie in Verwendung, später auch verschiedene Eigennamen, und heute gibt es bei der großen Anzahl neu entdeckter und gesicherter Planetoiden schon eine ganze Palette von Möglichkeiten der Benennung.

Entscheidend ist, dass es sich stets um eine *Ehrung, Würdigung, Erinnerung* oder einen anderen auszeichnungswürdigen und oft bemerkenswerten Aspekt (Person, Sache, . . .) handelt. *Man kann also diese Namen* **nicht** *kaufen.* Wohl aber kann man Vorschläge machen oder Ideen äußern, die dem CSBN (Committee for Small-Body Nomenclature) vorgelegt werden müssen. Über die Prozedur der dann folgenden Namenszuweisung geben wir im Anhang erstmals eine auszugsweise deutsche Übersetzung des in englischer Sprache vorliegenden amtlichen Dokumentes. Prinzipiell ist zu sagen, dass die von der Kommission 20 der IAU beschlossenen Regeln jederzeit geändert werden können. Der/Die Entdecker/in eines KP erhält das Privileg, innerhalb von zehn Jahren nach der Nummerierung des Asteroiden einen Namensvorschlag einzureichen, den das CSBN zu prüfen hat. Die auf diese Weise akzeptierten Namen werden dann als offiziell angesehen, wenn sie in den monatlich publizierten *Minor Planet Circulars* veröffentlicht worden sind.

Es ist in neuerer Zeit üblich geworden, dass auch andere Länder, nicht nur Österreich, Zusammenstellungen „ihrer" Namen vornehmen. So soll es auch hier wieder der Fall sein. Wir wollen also einen relativ weiten Rahmen stecken und den *Österreich-Bezug* wie folgt definieren:

Es werden Namen verzeichnet,

1) mit denen *Österreicher* die von ihnen entdeckten oder berechneten kleinen Planeten nach in- oder ausländischen Personen, Örtlichkeiten, allegorischen Bezeichnungen und anderen Motiven benannt haben,

2) die von anderen Entdeckern und Vorschlagsberechtigten für *österreichische* Personen, Orte, Ereignisse usw. gegeben wurden,

3) die auf Wunsch des CSBN kraft seiner Autorität für „Österreicher" vorgeschlagen wurden,

4) die für auswärtige Mitglieder der ÖAW, für die natürlich ebenso der Österreichbezug gilt, gegeben werden.

Es ist einsichtig, dass der Begriff „*Österreicher*" näher umschrieben werden muss. Bei den Personen handelt es sich um solche, die in Österreich geboren worden oder gestorben sind und/oder *einen wesentlichen Teil* ihres Lebens hier verbracht haben. Uns ist klar, dass es – besonders in der Zeit der Monarchie und des Überganges zur Republik – ambivalente Fälle geben kann, die wir nicht mehr oder gerade noch für Österreich reklamieren. Das ist oft eine schwere Entscheidung, und wir wollen damit sicher nicht Verdienste eines anderen Landes schmälern oder gar jemand beleidigen. In diesem Sinn bitten wir also um Verständnis für etwaige Ambiguitäten.

Bemerkenswert ist, dass viele kleine Planeten heute nicht mehr mühsam gesucht werden (wie etwa J. PALISA das tat), sondern durch automatische Teleskope oder jedenfalls mit dem höchsten Einsatz von Elektronik beobachtet werden. Wenn auch bestimmte Himmelsgegenden immer wieder aus verschiedenen Gründen systematisch untersucht werden, so sind doch die meisten Funde eigentlich „Zufallsentdeckungen", bei denen es dann oft Jahre dauert, bis sie gesichert und definitiv getauft werden können. Diesen Zeitunterschied zwischen Entdeckung (E) und Benennung (B) kann man aus unserer Arbeit unschwer ablesen und mit verschiedenen Umständen korrelieren.

Ein weiterer Aspekt unseres Verzeichnisses ist, wie schon im I. Teil ausgeführt, ein kulturhistorischer: Hier treten oft interessante Beweggründe für die Benennung zutage. Die fleißigsten Entdecker sind meist auch die häufigsten Benenner, wie etwa E. BOWELL (USA), F. BÖRNGEN und L. D. SCHMADEL (Deutschland), C. J. VAN HOUTEN und I. VAN HOUTEN-GROENEVELD (Holland), E. W. ELST (Belgien), M. TICHY und J. TICHA (Tschechien), um nur einige zu nennen. Manche von ihnen zeigen eine besondere Vorliebe für Österreich, seine Persönlichkeiten und seine Musik.

Es soll hier besonders erwähnt werden, dass gerade durch die erwähnte Möglichkeit, für einen entdeckten KP einen Namensvorschlag machen zu können, viele Amateure sich an der Beobachtung dieser Himmelskörper beteiligen. Die dabei aufgenommenen Bilder werden automatisch vermessen und die gewonnenen astrometrischen

Positionen oft direkt – über das Internet – an das Minor Planet Center gesendet. Dort werden alle Beobachtungen – von Profis wie auch von Amateurstationen – gesammelt, ausgewertet und zur Bestimmung bzw. Verbesserung der Bahnen der KP verwendet. Diese Entwicklung, zusammen mit der Zugänglichkeit und Vernetzung über das Internet, hat wesentlich zur Popularität der KP beigetragen. Aus der österreichischen Amateur-Szene sollen nur einige Namen erwähnt werden wie H. RAAB, E. MEYER, E. OBERMAIR und W. RIES, ohne die vielen anderen tüchtigen Beobachter vergessen zu wollen.

2. Die Liste der Österreich-Planeten

In dem nun folgenden Verzeichnis der Kleinplaneten-Namen werden – in gegen früher leicht veränderter Weise – gegeben:

Nummer und **Name** (in Fettdruck)

E: Entdeckungsdatum, Entdecker(in) und Ort der Entdeckung.

B: Person oder Institution, welche die Benennung durchgeführt hat, zugehörige Publikation (meist MPC) und Datum der Anzeige. Ist hier kein Name angegeben, so wird angenommen, dass der/die Erstentdecker/in für den Namen verantwortlich ist. In manchen Fällen ist auch die Person angeführt, die die folgende Würdigung („Citation") verfasst oder angeregt hat.

Citation: Im Gegensatz zu früher folgt jetzt die kurze Würdigung im vollen Wortlaut unverändert nach dem angeführten Dokument in *englischer* Sprache.

Allfällige Ergänzungen oder Korrekturen werden im nächsten Absatz in *deutscher* Sprache vorgenommen. Wir glauben, dass diese Zweisprachigkeit den Benutzern zumutbar ist. Es ist allerdings nicht mit so langen „Geschichten" zu rechnen, wie sie im Teil I dieser Arbeit enthalten sind; wir hoffen aber, dass diese zusätzlichen Informationen doch eine nützliche Anregung für die kulturhistorische Betrachtung bilden.

R: Hier werden ergänzende Angaben über die Quellen geboten, wobei die in Abschnitt 2.1 folgenden Abkürzungen verwendet werden.

Im Hauptverzeichnis werden die nach ihrer Nummer geordneten Namen angeführt, wobei diesmal auch die Mitglieder der ÖAW eingearbeitet sind. Zum Schluss folgen noch *alphabetische* Verzeichnisse aller behandelten Planeten mit beigefügter Nummer und das erwähnte offizielle Dokument über die Namensgebung.

2.1. Abkürzungen

ADB	Alte Deutsche Biographie	KP	Kleinplanet
AG	Astronomische Gesellschaft	MPC	Minor Planet Circular,
Alm.	Almanach		auch Minor Planet Center
AN	Astronomische Nachrichten	MPI	Max-Planck-Institut
Astrodat	Astronomendatei (HAUPT,	NDB	Neue Deutsche Biographie
	HOLL)	NR	Nachruf
B	Benennung	ÖAW	Österreichische Akademie
BdÖ	Das große Buch der		der Wissenschaften
	Österreicher	ÖBL	Österreichisches
Biogr.	Biographie		Biographisches Lexikon
DMPN	Dictionary of Minor Planets	PB	PALUZÍE-BORELL (1963)
	(SCHMADEL)	PG	Porträt-Galerie der AG
E	Entdeckung	R	Referenzen
EM	Ehrenmitglied	RI	Recheninstitut
H	HERGET (1955, 1968)	u. a.	unter anderem, auch und
IAU	Internationale		andere
	Astronomische Union	VJS	Vierteljahresschrift der AG
JB	Jahrbericht (eines Institutes)	W	WURZBACH, biographisches
k.M.	korrespondierendes Mitglied		Lexikon
	der ÖAW	w.M.	wirkliches Mitglied der
kM(A)	korrespondierendes Mitglied		ÖAW
	der ÖAW im Ausland		

2.2. Österreich-Planeten

(768) **Struveana**
  E: 1913 Oct. 04 by G. N. NEUJMIN at Simeis.
  B: DMPN: H 77.
Named in honor of FRIEDRICH GEORG WILHELM STRUVE (1793–1864), OTTO WILHELM STRUVE (1819–1905), directors of the Pulkowo Observatory and famous for their double star studies, HERMANN STRUVE (1854–1920), director of the Königsberg and Berlin Observatories (H 77).

FRIEDRICH G. W. STRUVE war deutscher Abstammung. Es gibt viele interessante Anekdoten über die Struves. Er war seit 1855 EM der ÖAW.
  R: NR Alm. d. ÖAW 15, 183 (1865).

(837) **Schwarzschilda**
  E: 1916 Sep. 23 by M. F. WOLF at Heidelberg.
  B: DMPN: H 82.
Named in honor of the German astronomer KARL SCHWARZSCHILD (1873–1916), director of the Göttingen (1901) and Potsdam (1909) Observatories. He worked in photometry, geometrical optics, stellar statistics and theoretical astrophysics (H 82).

KARL SCHWARZSCHILD war von 1896 bis 1899 Observator bei M. v. KUFFNER (KP 12586) und begann dort schon seine bahnbrechenden Untersuchungen über die Sternstrahlung. Er war dann Professor in Göttingen und Mitglied der Berliner Akademie.

R: NR VJS 58, 191 (1923) und andere.

(855) **Newcombia**

E: 1916 Apr. 03 by S. BELJAVSKIJ at Simeis.
B: DMPN: H 84.

Named in honor of the American astronomer SIMON NEWCOMB (1835–1909), professor of astronomy and director of the U.S. Nautical Almanac Office. NEWCOMB worked on cometary and planetary orbits and on the theory of the orbits of the Earth. He measured the velocity of light and determined the astronomical unit anew (H 84).

Benannt ist der Asteroid nach dem in den USA geborenen SIMON NEWCOMB, der seit 1904 kM(A) der ÖAW war.

R: NR Alm. d. ÖAW 60, 329 (1910).

(892) **Seeligeria**

E: 1918 May 31 by M. F. WOLF at Heidelberg.
B: DMPN: H 86.

Named in honor of the German astronomer HUGO HANS VON SEELIGER (1894–1924), director of the Munich Observatory. VON SEELIGER investigated the movement of the perihelion of Mercury and was one of the founders of stellar astronomy. He published books on the motion of binaries and on the theory of the heliometer (H 86).

Der Theoretiker HUGO VON SEELIGER war kM(A) der ÖAW seit 1895.

R: NR Alm. d. ÖAW 75, 209 (1925).

(999) **Zachia**

E: 1923 Aug. 09 by K. REINMUTH at Heidelberg.
B: DMPN: H 95.

Named in honor of the astronomer and mathematician FRANZ XAVER FREIHERR VON ZACH (1754–1832). He was the director of the Seeberg Observatory near Gotha, Germany. VON ZACH recovered planet (1) Ceres profiting the Gauss computings because the planet was lost in consequence of PIAZZI's serious illness. He founded the journal "Monatliche Correspondenz zur Beförderung der Erd- und Himmelskunde" which was the main medium for exchange and improvements in observation and data treatment of the first asteroids (H 95).

FRANZ XAVER FREIHERR VON ZACH (geboren in Pest, gestorben in Paris) war ein bekannter österreichisch-deutscher Astronom der Goethezeit, aber auch Geodät, Mathematiker, Wissenschaftshistoriker und Offizier. Er machte sich vor allem um die Erforschung des Sonnensystems und die Organisation der internationalen Astronomie verdient.

R: Über seine große Familie gibt es umfangreiche Forschungen durch Prof. P. BROSCHE, Bonn.

(1001) **Gaussia**
  E: 1923 Aug. 08 by S. BELJAVSKIJ at Simeis.
  B: DMPN: H 96.
Named in honor of CARL FRIEDRICH GAUSS (1777–1826), director of the Göttingen Observatory. With his new computing methods F. X. VON ZACH {see planet 999} rediscovered (1) Ceres (H 96).

JOHANN CARL FRIEDRICH GAUSS wurde geboren in Braunschweig und starb in Göttingen. Er war bedeutender Mathematiker und Astronom („Princeps mathematicorum"). Bei der Entdeckung und Wiederauffindung der ersten KP spielte er eine große Rolle. GAUSS war EM der ÖAW seit 1848.

R: NR Alm. d. ÖAW 6, 123 (1856).

(1024) **Hale**
  E: 1923 Dec. 02 by G. VAN BIESBROECK at Williams Bay.
  B: DMPN: H 98.
Named in honor of GEORGE ELLERY HALE (1868–1938), founder and first director of Yerkes and Mt. Wilson Observatories and inspirer of the observatory at Mt. Palomar. With J. E. KEELER {see planet 2261} he founded the "Astrophysical Journal", and he invented the spectroheliograph (H 98).

HALE war kM(A) seit 1910 und EM der ÖAW seit 1914.

R: NR Alm. d. ÖAW 88, 262 (1938), und weitere ausführliche Nachrufe und Biografien.

(1069) **Planckia**
  E: 1927 Jan 28 by M. F. WOLF at Heidelberg.
  B: DMPN: H 101.
Named in honor of the famous German physicist MAX KARL ERNST LUDWIG PLANCK (1858–1947), Nobel prize winner in 1918, on the occasion of his 80th birthday anniversary. He was a professor of

physics at Berlin University and the discoverer of the quantum nature of radiation (H 101).

PLANCK war kM(A) der ÖAW seit 1915, EM seit 1922.
R: NR Alm. d. ÖAW 98, 222 (1948).

(1123) Shapleya

E: 1928 Sep. 21 by G. N. NEUJMIN at Simeis.
B: DMPN: H 105.

Named by the discoverer (RI 509) in honor of the American astronomer and director of Harvard Observatory HARLOW SHAPLEY (1885–1972) (H 105).

HARLOW SHAPLEY war durch seine Arbeiten über Veränderliche Sterne und über die Milchstraße sehr bekannt. SHAPLEY war kM(A) der ÖAW seit 1936.
R: NR Alm. d. ÖAW 123, 315 (1973).

(1551) Argelander

E: 1938 Feb. 24 by Y. VÄISÄLÄ at Turku.
B: DMPN: MPC 2278.

Named in honor of F. W. A. ARGELANDER (1799–1875), director of the ancient observatory of Turku (Abo), and later director of the Bonn Observatory, and author of the famous "Bonner Durchmusterung".

ARGELANDER (geboren in der Nähe von Memel, gestorben in Bonn) war kM(A) der ÖAW seit 1851, EM seit 1872.
R: NR Alm. d. ÖAW 25, 234 (1872).

(1561) Fricke

E: 1941 Feb. 15 by K. REINMUTH at Heidelberg.
B: MPC 3930, issue date not found!

Named in honor of WALTER FRICKE, director of the Astronomisches Rechen-Institut in Heidelberg since 1955. The principal author of the FK4, he has also worked extensively on the system of astronomical constants. He served as president of IAU commissions 4 (1958–1964) and 8 (1970–1973) and vice president of the IAU (1964–1967).

Prof. FRICKE (geboren in Merseburg, gestorben in Heidelberg) war in seiner Funktion auch für Österreich stets hilfreich in Rat und Tat. Er war auch kM(A) der ÖAW.
R: NR Alm. d. ÖAW 138, 377 (1988).

(1631) Kopff

E: 1926 Oct. 05 by K. REINMUTH at Heidelberg.
B: MPC 3931, issue date not found!

Named in memory of AUGUST KOPFF (1882–1960), who as WOLF's assistant in Heidelberg discovered and observed many minor planets. In 1924 he became director of the Astronomisches Rechen-Institut in Berlin, and after the western section moved to Heidelberg he also became director of Heidelberg-Königstuhl observatory. He was responsible for constructing the FK3 and initiated work on the FK4.

AUGUST KOPFF (geboren und gestorben in Heidelberg) war Entdecker vieler Planetoiden, die er dann im Recheninstitut verwaltete. Er war kM(A) der ÖAW seit 1939.
    R: NR Alm. d. ÖAW 110, 469 (1960).

(1650) Heckmann
    E: 1937 Oct. 11 by K. REINMUTH at Heidelberg.
    B: MPC 3932, issue date not found!
Named in honor of OTTO HECKMANN, director of the Hamburg-Bergedorf Observatory from 1941 to 1962 and subsequently the first director of the European Southern Observatory, the foundation of which had been initiated by him. His research activities cover cosmology and several aspects of fundamental astronomy. He was president of the IAU from 1967 to 1970.

Prof. HECKMANN wurde 1901 in Opladen geboren und starb 1983 in Regensburg. Er war kM(A) der ÖAW seit 1962.
    R: NR Alm. d. ÖAW 133, 353 (1983).

(1677) Tycho Brahe
    E: 1940 Sep. 02 by Y. VÄISÄLÄ at Turku.
    B: MPC 4236, issue date not found!
Named for the great Danish-born astronomer TYCHO BRAHE (1546–1601).

TYCHO BRAHE war der Vorgänger KEPLERS am Prager Hof (vgl. KP 44613 Rudolf).

(1688) Wilkens
    E: 1951 Mar. 03 by M. ITZIGSOHN at La Plata.
    B: MPC 5449, 1980 Aug. 01.
Named in memory of ALEXANDER WILKENS, researcher in many branches of astronomy, most notably celestial mechanics. He worked for many years in Germany, then at the La Plata Observatory, where he produced two generations of celestial mechanicians before returning to his native country.

A. WILKENS war von 1904 bis 1905 Assistent an der Kuffner-Sternwarte (KP 12568) in Wien, bevor er wieder nach Deutschland bzw. dann nach La Plata ging.

(1693) **Hertzsprung**

E: 1935 May 05 by H. VAN GENT at Johannesburg.

B: DMPN: MPC 2822, issue date not found!

Named in honor of the late Prof. E. HERTZSPRUNG (1873–1967), who was Director of the Leiden Observatory from 1934 till 1945. A well-known authority in the field of astronomical photometry, he initiated the Leiden Variable Star Survey of the southern Milky Way, during which survey many asteroids and some comets were found.

EJNAR HERTZSPRUNG (geboren in Frederiksborg, gestorben in Tølløse) war ein dänischer Astronom. Er war auch kM(A) der ÖAW seit 1947.

R: NR Alm. d. ÖAW 119, 326 (1969).

(1904) **Massevitch**

E: 1972 May 09 by T. SMIRNOVA at the Crimean Astrophysical Observatory.

B: MPC 3936, issue date not found!

Honoring Dr. ALLA GENRICHOVNA MASSEVITCH, a well-known woman astronomer-astrophysicist, vice-president of the astronomical council of the USSR academy of sciences, the organizer of optical tracking of artificial earth satellites in the USSR.

ALLA MASSEVITCH (geboren 1918 in Tiflis) ist Mitglied mehrerer Akademien, darunter auch kM(A) der ÖAW seit 1985. Durch ihre Tätigkeit und ihre Reisen ist sie auch in Österreich gut bekannt.

(1905) **Ambartsumian**

E: 1972 May 14 by T. SMIRNOVA at the Crimean Astrophysical Observatory.

B: MPC 3937, issue date not found!

Named in honor of VIKTOR AMAZASPOVICH AMBARTSUMIAN, a world renowned scientist, the founder of the soviet school for astrophysics, president of the academy of sciences of the Armenian SSR, served as president of the IAU in 1961–64, director of the Byurakan Astrophysical Observatory.

Prof. AMBARTSUMIAN (geboren 1908 in Tiflis, Georgien, gestorben 1996 in Bjurakan, Armenien) war ein guter Freund Österreichs und oft bei uns auf Besuch. Er wurde 1956 kM(A) der ÖAW.

R: NR Alm. d. ÖAW 147, 497 (1997).

(1991) **Darwin**

E: 1967 May 06 by C. U. CESCO and A. R. KLEMOLA at the Yale-Columbia Southern Station, El Leoncito.

B: MPC 5282, 1980 Apr. 01.

Named in memory of CHARLES DARWIN (1809–1882), the English naturalist who first established the theory of organic evolution; much of his research was done in Argentina, and he crossed the Andes at a pass located some 100 km south of El Leoncito. This planet also honors his second son, GEORGE DARWIN (1845–1912), the astronomer noted for his pioneering application of detailed dynamical analyses to problems of cosmogony and geology.

Sir GEORGE H. DARWIN war kM(A) der ÖAW seit 1908.

R: NR Alm. d. ÖAW 63, 388 (1913).

(2001) **Einstein**

E: 1973 Mar. 05 by P. WILD at Zimmerwald.

B: MPC 4237, issue date not found!

Named in memory of ALBERT EINSTEIN (1879–1955), the greatest scientist of the twentieth century. Although EINSTEIN was born in Germany and died in the U.S., it is particularly fitting that his name be given to a Bernese minor planet, for he laid the foundations of his revolutionary scientific thoughts while working as an examiner in the Swiss patent office in Berne.

A. EINSTEIN war laut „Wissenschaftliche Nachrichten" von 1911–12 Österreichischer Staatsbürger.

R: Zahlreiche Nachrufe und Biogr.

(2021) **Poincaré**

E: 1936 June 26 by L. BOYER at Algiers.

B: MPC 4420, issue date not found!

Named in honor of HENRI POINCARÉ (1854–1912), distinguished French mathematician and celestial mechanician, contributor to the theory of functions, modern algebra, algebraic topology, number theory, and les methodes nouvelles de la mecanique celeste.

JULES HENRI POINCARÉ (geboren in Nancy, gestorben in Paris) war nicht nur ein bedeutender französischer Mathematiker, sondern auch Philosoph. Er war kM(A) der ÖAW seit 1903, EM seit 1908.

R: NR Alm. d. ÖAW 63, 383 (1913).

(2069) **Hubble**

E: 1955 Mar. 29 at the Goethe Link Observatory, Indiana University.

B: N. U. MAYALL, MPC 8403, 1983 Dec. 20.

Named in memory of EDWIN P. HUBBLE (1889–1953), who provided the first comprehensive exploration of the universe beyond our own galaxy. He established a self-consistent distance scale as far as the 2.5-m Mount Wilson reflector could reach, and his classification scheme for galaxies is still the standard. He discovered the unique minor planet (1373) Cincinnati. His greatest achievement, however, known as Hubble's law of redshifts, can be interpreted as observational basis for the expanding universe. Name proposed by F. K. EDMONDSON. Citation written by N. U. MAYALL.

EDWIN POWELL HUBBLE (geboren in Missouri, gestorben in Kalifornien) studierte Physik und Astronomie in Chicago sowie Rechtswissenschaften in Oxford. Seine wichtigen Beiträge zur Kosmologie sind oben angegeben. Das Hubble-Weltraumteleskop (HST) wurde nach ihm benannt. HUBBLE war EM der ÖAW seit 1947.

R: NR Alm. d. ÖAW 104, 411 (1954).

(2288) Karolinum

E: 1979 Oct. 19 by L. BROZEK at the Klet' Observatory.

B: MPC 5525, 1980 Oct. 01.

Named for the original main building, still in use, of the Charles University, founded in Prague in 1348.

Die Karls-Universität Prag ist die älteste Universität Mitteleuropas und wurde vom römisch-deutschen Kaiser KARL IV. gegründet.

(2947) Kippenhahn

E: 1955 Aug. 22 by I. GROENEVELD at Heidelberg.

B: MPC 18136, 1991 Apr. 28.

Named in honor of RUDOLF KIPPENHAHN (1926–), German astronomer, director of the Max-Planck-Institut für Physik und Astrophysik at Garching and currently a vice president of the IAU.

Prof. KIPPENHAHN, geboren in Bärringen, Böhmen, ist Mitglied mehrerer Akademien. In seinem Ruhestand ist er Verfasser viel gelesener astronomischer Bücher. Er ist kM(A) der ÖAW seit 1995.

(3103) Eger

E: 1982 Jan. 20 by M. LOVAS at Piszkéstető.

B: MPC 23135, 1994 Feb. 26.

Named for a famous Hungarian city. In 1762 its bishop, Count KAROLY ESTERHÁZY, established a university there at his own expense. The university included an observatory, which he furnished on consultation with MAXIMILIAN HELL in Vienna, instruments being made both in Vienna and London. The Eger observations began in 1778, and the results were published in HELL's annual astronomical ephemerides. The city is also known for the successful stand of its populace against the besieging Turks in 1552. It is situated in an excellent wine-growing area and is known internationally for its hearty red Egri Bikaver ("bull's blood").

Die enge Verbindung mit Österreichs Astronomen und deren Hilfe bei der Einrichtung des Observatoriums rechtfertigt die Aufnahme in unsere Liste. – Ferner ist Eger für seine Thermalquelle und das Thermalbad bekannt.

(3184) **Raab**

E: 1949 Aug. 22 by E. L. JOHNSON at Johannesburg.
B: B. G. MARSDEN and G. V. WILLIAMS, MPC 27124, 1996 May 03.
Named in honor of HERBERT RAAB (b. 1969), author of the widely-used and acclaimed Astrometrica software package. Astrometrica has enabled many amateur astronomers to participate in their own astrometric programs on comets and minor planets. RAAB is a software developer and is also president of the Linzer Astronomische Gemeinschaft, the oldest amateur association in Austria. Name proposed by B. G. MARSDEN and G. V. WILLIAMS.

HERBERT RAAB ist weltweit bekannt durch die Entwicklung und ständige Verbesserung des Computerprogrammes „*Astrometrica*", mit dessen Hilfe CCD-Aufnahmen von Asteroiden vermessen werden können. Amateurastronomen, aber auch viele Profis, verwenden diese Software. RAAB ist begeisterter Amateurastronom und Mitglied der Linzer Astronomischen Gemeinschaft (siehe KP 7491 Linzerag), sowie Consultant der IAU Kommission 20 (Minor Planets and Comets). Die Namensgebung dieses KP erfolgte auf Vorschlag des Direktors des Minor Planet Center.

(3282) **Spencer Jones**

E: 1949 Feb. 19 at the Goethe Link Observatory, Indiana University.
B: F. K. EDMONDSON and J. S. TENN, MPC 16041, 1990 Mar. 11.
Named in memory of HAROLD SPENCER JONES (1890–1960), successively astronomical assistant at the Royal Greenwich Observatory, H.M. astronomer at the Cape of Good Hope, and Astronomer Royal (1933–1955). He also served as president of the IAU (1945–1948). His work was devoted to fundamental positional astronomy, and he conclusively demonstrated that the small residuals in the apparent motions of the planets were due to the irregular rotation of the earth. He led the worldwide effort to determine the length of the

astronomical unit by triangulating the distance to (433) Eros when it passed near the earth in 1930–31. Name proposed by F. K. EDMONDSON. Citation prepared by J. S. TENN.

Der Astronomer Royal war ab 1950 EM der ÖAW. Sein Familienname war „Spencer Jones".

R: NR bei der ÖAW offenbar nicht vorhanden.

(3366) **Gödel**
E: 1985 Sep. 22 by T. SCHILDKNECHT at Zimmerwald.
B: MPC 27125, 1996 May 03.

Named for KURT GÖDEL (1906–1978), American logician of Austrian origin. He is the author of the famous article Über formal unentscheidbare Sätze der Principia Mathematica und verwandter Systeme (1931). His ideas accompany the discoverer through many observing nights.

KURT GÖDEL (geboren 1906 in Brünn, gestorben 1978 in Princeton, New Jersey) war Mathematiker und Logiker. GÖDEL wird von vielen als der bedeutendste Logiker des 20. Jahrhunderts angesehen. Er war auch Mitglied des Wiener Kreises.

(3386) **Klementinum**
E: 1980 Mar. 16 by L. BROZEK at Klet'.
B: MPC 21130, 1992 May 16.

Named for the Jesuit college of St. Clemens in the Old Town of Prague. At the beginning of the eighteenth century a mathematical museum and astronomical observatory were established there, the observation tower being completed in 1722. Klementinum was long an important center for astronomical, meteorological and geophysical research and measurements. Now it contains the national and university libraries.

(3433) **Fehrenbach**
E: 1963 Oct. 15 at the Goethe Link Observatory, Indiana University.
B: F. K. EDMONDSON, MPC 17027, 1990 Oct. 04.

Named in honor of CHARLES FEHRENBACH, who pioneered the successful use of the objective prism to measure stellar radial velocities. One of his instruments was used to identify members of the Magellanic Clouds from their radial velocities, starting in 1961 during the ESO site survey in southern Africa and continuing in 1968 on La Silla in Chile. He served as vice president of the IAU from 1973 to 1979, as a member of ESO and president of its Commission on Instruments, and as a member and president of the council of the

Canada-France-Hawaii Observatory. His many honors include membership in the French Academy of Sciences, the gold medal of the CNRS and the grand scientific prize of the city of Paris. Name proposed by F. K. EDMONDSON.

CHARLES FEHRENBACH (geboren 1914 in Straßburg, gestorben 2008 in Nîmes) war Mitglied der Académie française. Er war Direktor des Observatoire de Haute-Provence (OHP) bis zum Jahre 1983 und dann dessen Ehrendirektor sowie emeritierter Professor der Université de Haute-Provence (Marseille). Er war kM(A) der ÖAW und gehörte auch zahlreichen anderen Akademien an. Durch Einladungen hat er 15 Jahre lang Beobachtungen der Grazer Astronomen am OHP ermöglicht.
R: NR Alm. d. ÖAW 158 (2009).

(3484) Neugebauer
E: 1978 July 10 by E. F. HELIN and E. M. SHOEMAKER at Palomar.
B: MPC 14632, 1989 May 20.
Named in honor of the NEUGEBAUERS, prominent family of physicists and mathematicians who have each made significant contributions in their chosen fields. GERRY NEUGEBAUER is chairman of the division of physics, mathematics and astronomy, California Instititute of Technology, and director of Palomar Observatory; MARCIA NEUGEBAUER is project scientist for the Comet Rendezvous Asteroid Flyby mission at the Jet Propulsion Laboratory; and OTTO E. NEUGEBAUER, on the faculty of the Institute for Advanced Study, Princeton, was professor emeritus, Brown University, and celebrated his ninetieth birthday on 1989 May 26.

OTTO NEUGEBAUER (1899–1990) war kM(A) seit 1970 und EM der ÖAW seit 1987. K. FERRARI (siehe KP 7146 Konradin) hat ihm, „dem großen Erforscher der antiken Sternkunde", die 2. und 3. Auflage seines Buches „Der Stern von Bethlehem in astronomischer Sicht" gewidmet. NEUGEBAUER wurde in Innsbruck geboren, maturierte und studierte in Graz, München und Göttingen. Als Gegner des Nationalsozialismus legte er 1932 sein Amt nieder, ging als Gastprofessor nach Kopenhagen und 1939 an die Brown University in Providence (USA), wo er bis zu seiner Emeritierung blieb. Er hat bedeutende Werke über die antike Mathematik und Astronomie geschrieben.
R: NR in Alm. d. ÖAW 140, 381 (1990).

(3727) Maxhell
E: 1981 Aug. 07 by A. MRKOS at Klet'.
B: MPC 26424, 1996 Jan. 05.

Named in memory of MAXIMILIAN HELL (1720–1792), famous for his determination of the solar parallax from his observations of the transit of Venus in 1769. Appointed director of the imperial observatory in Vienna in 1755, he prepared and published an important series of astronomical ephemerides. Name suggested by astronomers at the Astronomical Institute at Tatranská Lomnica.

Der in Schemnitz (Ungarn) geborene Pater MAXIMILIAN HELL war von 1752 bis 1755 Direktor der Sternwarte Klausenburg, dann bis 1792 erster Direktor der im Gebäude der heutigen ÖAW gegründeten Wiener Sternwarte. Seine Expedition zum Venusdurchgang auf die Insel Wardoe (1768) stand lange Zeit im Mittelpunkt heftiger Kritik. Erst durch S. NEWCOMB (1883) wurden seine Ergebnisse rehabilitiert und anerkannt.

R: Zahlreiche Nachrufe und Lebensbeschreibungen, u. a. NDB 8, 473 (1969).

(3791) **Marci**

E: 1981 Nov. 17 by A. MRKOS at Klet'.

B: J. TICHÁ and M. ŠOLC, MPC 27733, 1996 Aug. 28.

Named in memory of JAN MARCUS MARCI of Kronland (1595–1667), Czech physicist, mathematician, astronomer and physician at the Charles University in Prague during the Thirty Years War. Twenty years before NEWTON, MARCI thoroughly described the spectral dispersion and diffraction of light, color effects on thin layers and rainbow colors. He also studied elastic and inelastic collisions of spheres, the motion of a pendulum and tried to solve the problem of squaring the circle. Name suggested by J. TICHÁ and M. ŠOLC.

(3847) **Šindel**

E: 1982 Feb. 16 by A. MRKOS at Klet'.

B: J. TICHÁ and M. ŠOLC, MPC 27733, 1996 Aug. 28.

Named in memory of JAN ONDŘEJOV, known as ŠINDEL (1375–1456), Czech medieval astronomer, mathematician, physician, professor at the universities in Nuremberg, Vienna and Prague, rector of the Charles University in Prague. He designed the famous Prague astronomical clock, which was built by the clockmaster NICOLAS OF KADAŇ in 1410 and is still working today. Name suggested by J. TICHÁ and M. ŠOLC.

(3887) **Gerstner**

E: 1985 Aug. 22 by A. MRKOS at Klet'.

B: J. TICHÁ and M. ŠOLC, MPC 27733, 1996 Aug. 28.

Named in memory of the physicist FRANTIŠEK JOSEF GERSTNER (1756–1832) and his son FRANTIŠEK ANTONÍN GERSTNER (1795–1840). F. J. GERSTNER worked at the observatories in Vienna and Prague but then turned from astronomy to applied mathematics, physics, mechanics and metallurgy. He founded the Prague Polytechnic School, later Technical University, in 1806. He also designed the horse railroad from České Budějovice to Linz. F. A. GERSTNER continued building this horse railroad, which was completed in 1832 and is considered the first railroad in continental Europe. Name suggested by J. TICHÁ and M. ŠOLC.

1781 ging er nach Wien, um Medizin zu studieren, brach jedoch das Studium ab und arbeitete auf der Wiener Sternwarte. 1784, nach dreijährigem Praktikum, wurde er Adjunkt des Professors STRNAD an der Sternwarte Prag. 1785 publizierte er seine erste astronomische Arbeit, in der er die geografische Länge einer Reihe europäischer Städte korrigierte. Als Anerkennung seiner Arbeit ernannte ihn die Königliche Böhmische Gesellschaft der Wissenschaften (Královská česká společnost nauk) zu einem ordentlichen Mitglied. Auch seine weiteren Werke im Bereich der Astronomie brachten ihm Anerkennung in den europäischen wissenschaftlichen Kreisen.
  R: Biographie ÖBL 1, 430 (1957).

(3905) **Doppler**
  E: 1984 Aug. 28 by A. MRKOS at Klet'.
  B: J. TICHÁ and M. ŠOLC, MPC 27734, 1996 Aug. 28.
Named in memory of the Austrian physicist CHRISTIAN DOPPLER (1803–1853), professor of mathematics and geometry at the Technical University in Prague and later the first director of the Physical Institute in Vienna. The well-known effect relating a shift in observed wavelength to the relative motion of source and observer was formulated by DOPPLER in Prague in his book "Ueber das farbige Licht der Doppelsterne" (1842). Name suggested by J. TICHÁ and M. ŠOLC.
  R: Zahlreiche Nachrufe und Lebensbeschreibungen; u. a. ÖBL 1, 196 (1957), NDB 4, 76 (1959), W 3, 370 (1857).

(3949) **Mach**
  E: 1985 Oct. 20 by A. MRKOS at Klet'.
  B: J. TICHÁ and M. ŠOLC, MPC 27734, 1996 Aug. 28.
Named in memory of ERNST MACH (1838–1916), professor of physics at the universities of Graz and Prague and later professor of philosophy in Vienna. He made investigations on supersonic motion, explosions, electric sparks and philosophical positivism, and the

Mach number and Mach's principle are forever associated with his name. Name suggested by J. TICHÁ and M. ŠOLC.

Der in Mähren Geborene und in München Gestorbene war ab 1867 k.M. und ab 1870 w.M. der ÖAW und Sekretär der math.-nat. Klasse 1897–98.

R: NR Alm. d. ÖAW 66, 328 (1916); BdÖ 309.

(3979) **Brorsen**

E: 1983 Nov. 08 by A. MRKOS at Klet'.

B: J. TICHÁ, MPC 27734, 1996 Aug. 28.

Named in memory of THEODOR BRORSEN (1819–1895), Danish astronomer, known for his discoveries of five comets and his studies of the gegenschein. After studying and working in Kiel, Heidelberg and Altona he worked at the private observatory of baron JOHN PARISH in Senftenberg (Žamberk) in eastern Bohemia from 1847 to 1870. Name suggested by J. TICHÁ.

1846 arbeitete BRORSEN am astronomischen Observatorium in Kiel, 1847 in Altona. Das Angebot einer Stelle am Observation „Runde Tårn" (Runder Turm) in Kopenhagen lehnte er ab. Stattdessen nahm er eine Stelle am Privatobservatorium des Baron JOHN PARISH in Senftenberg in Böhmen an. 1854 bewarb sich BRORSEN ohne Erfolg auf die frei gewordene Direktorenstelle des Observatoriums in Altona. Nach dem Tode des Barons PARISH 1858 ließen dessen Erben das Observatorium in Senftenberg abreißen und die Instrumente verkaufen, obwohl BRORSEN angeboten hatte, gratis weiterzuarbeiten. Trotzdem blieb BRORSEN noch zwölf weitere Jahre in Senftenberg (siehe KP 59001) und setzte die Beobachtungen mit seinen eigenen Instrumenten fort.

(4062) **Schiaparelli**

E: 1989 Jan. 28 at the Osservatorio San Vittore.

B: MPC 15090, 1989 Sep. 15.

Named in memory of GIOVANNI VIRGILIO SCHIAPARELLI (1835–1910), discoverer of the connection between comets and meteor streams. A great observer of Mars, he discovered the famous "canali" and drew some fine maps. He made measurements of double stars and of the rotational periods of Mercury and Venus, and he discovered (69) Hesperia. From 1862 to 1900 he was director of the Brera Observatory in Milan.

Er war kM(A) der ÖAW ab 1875, EM seit 1893.

R: NR Alm. d. ÖAW 61, 378 (1911).

(4158) Santini

E: 1989 Jan. 28 at the Osservatorio San Vittore.

B: MPC 16043, 1990 Mar. 11.

Named in memory of GIOVANNI SANTINI (1786–1877), director of the Padua Observatory from 1817 to 1867, a great observer of minor planets and comets and an indefatigable computer of their orbits and perturbations. His two-volume textbook "Elementi di Astronomia" was used by virtually all the Italian astronomers of the nineteenth century.

SANTINI war w.M. der ÖAW ab 1847, dann kM(A) im Ausland seit 1867.

R: NR Alm. d. ÖAW 28, 159 (1878).

(4297) Eichhorn

E: 1938 Apr. 19 by W. DIECKVOSS at Bergedorf.

B: A. R. UPGREN, MPC 28621, 1996 Dec. 24.

Named in honor of HEINRICH KARL EICHHORN (b. 1927), Austrian-American astronomer, educator and scholar, innovator in the astronomy of stellar positions and motions. He developed the central-overlap and other astrometric reduction methods that greatly improve their rigor, increasing the precision of stellar parallaxes and proper motions and thus also of the distance scale of the universe. Name proposed by A. R. UPGREN.

HEINZ EICHHORN war geborener Wiener und studierte bei A. PREY, K. GRAFF und J. RADON. Nach mehreren Auslandsaufenthalten in England und den USA wurde er schließlich Professor bzw. Institutsvorstand an der University of South Florida (Tampa) und später der University of Florida in Gainesville. Er blieb seiner Heimat zeitlebens verbunden, wurde Honorarprofessor in Graz und Wien und 1987 k.M. der ÖAW. Er starb 1999 noch vor der großen Sonnenfinsternis in Österreich und vor seinem Goldenen Doktorjubiläum.

R: NR Alm. d. ÖAW 149, 431 (1999) und weitere Nachrufe.

(4385) Elsässer

E: 1960 Sept. 24 by C. J. VAN HOUTEN and I. VAN HOUTEN-GROENEVELD at Leiden on Palomar Schmidt plates taken by T. GEHRELS.

B: L. D. SCHMADEL, MPC 18141, 1991 Apr. 28.

Named in honor of HANS F. ELSÄSSER (1929–2003), professor of astronomy at Heidelberg University and since 1968 first director of the Max-Planck-Institut für Astronomie. During 1962–1975 he was also director of the Heidelberg Observatory at Königstuhl. A scientific member of the Max-Planck-Gesellschaft, ELSÄSSER was a founder of MPIA in Heidelberg and its associated observatory at Calar Alto. He has made important contributions to the study of interplanetary matter and the zodiacal light, the optics of the earth's atmosphere, the structure of the Galaxy and the Magellanic Clouds and star formation. He was deeply concerned with the design and establishment of large telescopes and their auxiliary instrumentation at Calar Alto. ELSÄSSER served as president of the IAU Commission 21 during 1970–1973. His many honors include membership in the scientific academies of Austria, Halle (Leopoldina) and Heidelberg. Name proposed and citation prepared by L. D. SCHMADEL.

Prof. Dr. HANS FRIEDRICH ELSÄSSER (geboren 1929 in Aalen, gestorben 2003 in Heidelberg) war kM(A) der ÖAW seit 1981.

R: NR Alm. d. ÖAW 153, 471 (2003).

(4386) **Lüst**

E: 1960 Sept. 26 by C. J. VAN HOUTEN and I. VAN HOUTEN-GROENEVELD at Leiden on Palomar Schmidt plates taken by T. GEHRELS.

B: MPC 18141, 1991 Apr. 28.

Named in honor of REIMAR LÜST (1923–), German astronomer, former director general of the Max-Planck-Gesellschaft and of the European Space Agency.

REIMAR LÜST, geboren in Wuppertal, vielfacher Ehrendoktor und Mitglied zahlreicher Akademien, hat sich als Generaldirektor der ESA auch für österreichische Belange immer wieder eingesetzt. Er ist kM(A) der ÖAW seit 1980.

(4674) **Pauling**

E: 1989 May 02 by E. F. HELIN at Palomar.

B: Caltech, MPC 17981, 1991 Mar. 30.

Named in honor of Professor LINUS PAULING on the occasion of his ninetieth birthday, 1991 Feb. 28. PAULING has had a long and distinguished career, spending 37 years as a Caltech faculty member, including 22 years as chairman of Caltech's Division of Chemistry and Chemical Engineering. He is the recipient of Nobel prizes for both Chemistry and Peace. The discoverer, and her husband RONALD, a Caltech graduate, are long-time admirers of PAULING. Asteroid tribute endorsed by the Caltech community.

LINUS CARL PAULING (geboren 1901 in Portland, Oregon; gestorben 1994 in Big Sur, Kalifornien) war ein US-amerikanischer Chemiker deutscher Abstammung. Er erhielt 1954 den Nobelpreis für Chemie und 1962 den Friedensnobelpreis als besondere Auszeichnung für seinen Einsatz gegen Atomwaffentests und ist damit neben MARIE CURIE der einzige Nobelpreisträger in unterschiedlichen Kategorien. Er war kM(A) der ÖAW seit 1961.

R: NR Alm. d. ÖAW 145, 477 (1994).

(4804) Pasteur

E: 1989 Dec. 02 by E. W. ELST at the European Southern Observatory.
B: MPC 19340, 1991 Nov. 21.

Named in memory of the great French chemist and microbiologist LOUIS PASTEUR (1822–1895), who proved that fermentation and disease are caused by micro-organisms. His invention of the principle of immunization was successfully applied for the first time against rabies in 1885. In 1888 the celebrated Pasteur Institute was established in Paris, and the process of pasteurization is well known throughout the whole world.

Der Franzose LOUIS PASTEUR war kM(A) der ÖAW seit 1882, EM seit 1893.

R: NR Alm. d. ÖAW 46, 82 (1896).

(4991) Hansuess

E: 1981 Mar. 01 by S. J. BUS at Siding Spring in the course of the U.K. Schmidt-Caltech Asteroid Survey.
B: MPC 41382, 2000 Oct. 13.

HANS SUESS (1909–1993) was one of the pioneers of cosmochemistry. He proposed that the relative abundance of each element depends on its mass and that patterns in elemental abundances were caused by a combination of nuclear properties and the mechanisms by which heavy elements are created in stars.

HANS EDUARD SUESS, geboren in Wien, gestorben in San Diego, wurde 1935 an der Universität Wien promoviert und emigrierte 1950 in die USA. Er war seit 1967 kM(A) der ÖAW.

R: NR Alm. d. ÖAW 144, 367 (1994).

(5803) Ötzi

E: 1984 July 21 by A. MRKOS at Klet'.
B: M. TICHÝ, MPC 40701, 2000 May 23.

Ötzi, or Iceman, is a popular name for a prehistoric man of the late Stone Age. His mummified body was found on the Similaun Glacier in the Tirolean Ötztal Alps, on the Italian-Austrian border in 1991. The name was proposed by M. TICHÝ.

Der Mann vom Hauslabjoch, allgemein bekannt als Ötzi, ist eine Gletschermumie aus der ausgehenden Jungsteinzeit (Neolithikum) bzw. der Kupferzeit (Eneolithikum, Chalkolithikum). Am 19. September 1991 wurde die etwa 5300 Jahre alte Mumie beim Tisenjoch nahe dem Hauslabjoch in den Ötztaler Alpen oberhalb des Niederjochferners in 3210 m Höhe gefunden. Die offizielle Bergung wurde am 23. September 1991 durch den damaligen Vorstand des Instituts für Gerichtsmedizin der Universität Innsbruck, Prof. RAINER HENN, durchgeführt.

Alles Weitere findet sich im Südtiroler Archäologiemuseum in Bozen.

(6044) **Hammer-Purgstall**

E: 1991 Sep. 13 by L. D. SCHMADEL and F. BÖRNGEN at Tautenburg.

B: L. D. SCHMADEL (auf Anregung des Präsidiums der ÖAW), MPC 30798, 1997 Oct. 16.

Named in memory of JOSEPH FREIHERR VON HAMMER-PURGSTALL (1774–1856) on the occasion of the 150th anniversary of the Austrian Academy of Sciences. As a profound orientalist, poet and historian, he was the founder and first president (1847–1849) of the (then Imperial) Academy of Sciences in Vienna. His most famous work is the ten-volume History of the Osman Empire. Name proposed by the first discoverer following a suggestion from the presidency of the Austrian Academy of Sciences, citation prepared by H. HAUPT.

FREIHERR VON HAMMER-PURGSTALL wurde 1774 in Graz geboren, gestorben ist er 1856 in Wien.

R: NR. Alm. d. ÖAW 8, 71 (1858); BdÖ 171.

(6079) **Gerokurat**

E: 1981 Feb. 28 by S. J. BUS at Siding Spring in the course of the U.K. Schmidt-Caltech Asteroid Survey.

B: MPC 46007, 2002 June 24.

GERO KURAT (b. 1938), of the Natural History Museum in Vienna, is curator of the Vienna meteorite collection and president of the Meteoritical Society. His provocative ideas on the origin of meteorites have caused scientists to question basic paradigms about the origin of our solar system.

GERO KURAT wurde in Klagenfurt geboren. Studium der Mineralogie und Petrografie mit Nebenfach Physik an der Universität Wien, am

Naturhistorischen Museum in Wien (Mineralogisch-Petrografische Abteilung) tätig und ab 1968 Abteilungsleiter, seit 1. Jänner 2004 im Ruhestand. Das wissenschaftliche Interesse konzentriert sich auf Meteorite, Mikrometeorite (kosmischen Staub), Mondgesteine, Gesteine des Erdmantels und die unbemannte Erkundung planetarer Oberflächen.

KURAT war auch Professor an der Universität Wien, 1993 k.M. und seit 1995 w.M. der ÖAW.

(6145) Riemenschneider

E: 1960 Sep. 26 by C. J. VAN HOUTEN and I. VAN HOUTEN-GROENEVELD on Palomar Schmidt plates taken by T. GEHRELS.

B: MPC 26764, 1996 Mar. 05.

Named for TILMAN RIEMENSCHNEIDER (c.1460–1531), German sculptor in stone and in wood. He lived in Würzburg and served as the city's mayor. During the "Bauernkrieg" of 1525 he was on the side of the peasants, who lost the war. This probably resulted in his being tortured, and there is no record that he continued to sculpt after that time. In southern Germany and Austria there exist many wonderful altars from his hands, notably in Rothenburg, Creglingen, Heidelberg and Würzburg.

(6157) Prey

E: 1991 Sep. 09 by L. D. SCHMADEL and F. BÖRNGEN at Tautenburg.

B: L. D. SCHMADEL and H. HAUPT, MPC 31296, 1998 Feb. 11.

Named in memory of ADALBERT PREY (1873–1949), professor of astronomy in Innsbruck, Prague and, eventually, Vienna, where he was elected to membership in the Austrian Academy of Sciences. After World War II until his death he served as one of the two secretaries of the Academy. He worked on the motion of 70 Oph (sometimes called "Prey's star"), as well as on the dynamics of minor planets and the moon. He thoroughly investigated the field of isostasy and gravitation, to which he devoted a series of papers. Named by the first discoverer following a suggestion by H. HAUPT, who also prepared the citation.

R: NR Alm. d. ÖAW 100, 333 (1950), ÖBL 8, 272 (1982), NDB 20, 712 (2001).

(6231) Hundertwasser

E: 1985 Mar. 20 by A. MRKOS at Klet'.

B: J. TICHÁ, MPC 40701, 2000 May 23.

FRIEDENSREICH HUNDERTWASSER (FRIEDRICH STOWASSER, 1928–2000) was an Austrian graphic artist and painter whose decorative

abstract style follows in the Secessionist (Art Nouveau) tradition of
GUSTAV KLIMT and EGON SCHIELE. The name was suggested by J.
TICHÁ.

FRIEDRICH STOWASSER wurde 1928 in Wien geboren und starb 2000
an Bord der Queen Elizabeth 2. In Österreich gibt es einige berühmte
Gebäude, die von HUNDERTWASSER entworfen wurden (Hundertwas-
serhaus in Wien, Kirche in Bärnbach in der Steiermark, u. a.).
    R: Es gibt zahlreiche Referenzen und Abhandlungen über sein
Leben; BdÖ 212.

(6353) **Semper**

E: 1977 Oct. 16 by C. J. VAN HOUTEN and I. VAN HOUTEN-
GROENEVELD on Palomar Schmidt plates taken by T. GEHRELS.
    B: MPC 26766, 1996 Mar. 05.
Named for the German architect GOTTFRIED SEMPER (1803–1873).
On the recommendation of SCHINKEL he became a professor at
Dresden Academy. He built the Dresden museum and synagogue, as
well as the most famous "Semper Operahouse", destroyed during
World War II but rebuilt. SEMPER also worked in Zurich and Vienna,
where many buildings designed by him – including the Zurich
Observatory – are still in use.

In Wien baute er das Kunsthistorische und das Naturhistorische
Museum und die Neue Hofburg. Sterbedatum richtig 1879. SEMPER
war EM der ÖAW ab 1877.
    R: NR Alm. d. ÖAW 29, 136 (1879); BdÖ 497.

(6457) **Kremsmünster**

E: 1992 Sep. 02 by L. D. SCHMADEL and F. BÖRNGEN at Tautenburg.
    B: L. D. SCHMADEL, MPC 31610, 1998 Apr. 11.
Named as a celestial tribute to the famous Benedictine monastery,
educational center and observatory in Austria. A place of scientific
effort in many disciplines, the Kremsmünster Observatory remains
one of the most outstanding cultural institutions in Austria – even a
quarter of a millennium after its founding. This planet is also named
in memory of Abbot AUGUSTIN RESLHUBER (1808–1875), an ardent
observer to whom it was not granted to discover a minor planet
himself. Named by the first discoverer and citation prepared with the
support of Father AMAND KRAML.

Das Benediktinerstift Kremsmünster wurde im Jahre 777 durch den
Bayernherzog TASSILO III. gegründet (Tassilokelch). Ursprünglich
im Ostteil („Traungau") des Herzogtums Bayern liegend, gehörte der

Ort seit dem 12. Jahrhundert zum Herzogtum Österreich. Seit 1918 gehört der Ort zum Bundesland Oberösterreich.

Kremsmünster hat die längste meteorologische Beobachtungsreihe in Österreich und besitzt im „Astronomischen Turm" zahlreiche alte astronomische Geräte und Bilder.

R: Archiv der Sternwarte.

(6501) **Isonzo**

E: 1993 Dec. 05 at Farra d'Isonzo.
B: MPC 30099, 1997 June 20.

Named for the river near which the town of Farra d'Isonzo and its observatory are located. Long known as a landmark in north-eastern Italy, the river has been a crossing point to eastern Europe since Roman times.

Der Fluss ist in tragischer Erinnerung wegen der hier geführten 12 Schlachten zwischen Österreich und Italien im Ersten Weltkrieg, die trotz hoher Verluste auf beiden Seiten keine Kriegsentscheidung brachten.

(6597) **Kreil**

E: 1988 Jan. 09 by A. MRKOS at Klet'.
B: J. TICHÁ and M. ŠOLC, MPC 42355 2001 Mar. 09.

KARL KREIL (1798–1862), the sixth director of the Klementinum observatory in Prague, carried out geomagnetic measurements in the Austro-Hungarian empire. He was later director of the Central Institute for Meteorology and Geomagnetism in Vienna. The name was suggested by J. TICHÁ and M. ŠOLC.

KARL KREIL studierte zuerst in Kremsmünster, bevor er Assistent an der Sternwarte Wien wurde. Dann war er ab 1838 Adjunkt an der Sternwarte Prag und ab 1845 ihr Direktor, bevor er als Chef an die Zentralanstalt nach Wien wechselte. Er war w.M. der ÖAW seit 1847.

R: NR Alm. d. ÖAW 13, 118 (1863), ÖBL 4, 245 (1969), ADB 17, 101 (1883).

(6712) **Hornstein**

E: 1990 Feb. 23 by A. MRKOS at Klet'.
B: J. TICHÁ and M. ŠOLC, MPC 42356, 2001 Mar. 09.

KARL HORNSTEIN (1824–1882), the eighth director of the Klementinum observatory in Prague, studied the orbits of minor planets, including the irregularities in their distribution that were later called the Kirkwood gaps. The name was suggested by J. TICHÁ and M. ŠOLC.

HORNSTEIN war Assistent an der Wiener Sternwarte, dann Gymnasialprofessor in Wien, später Adjunkt, Dozent und Professor für Mathematik an der Universität Graz, ehe er zum Direktor der Sternwarte Prag avancierte. Er war k.M. der ÖAW seit 1857.
   R: NR Alm. d. ÖAW 33, 203 (1883), ÖBL 2, 425 (1959).

(6768) Mathiasbraun
   E: 1983 Sep. 07 by A. MRKOS at Klet'.
   B: J. TICHÁ and M. TICHÝ, MPC 40701, 2000 May 23.

MATTHIAS (MATYÁŠ) BERNARD BRAUN (1684–1738) was the most prominent sculptor of the Baroque period in Bohemia. His well-known works are allegories of virtues and vices at the Kuks Castle. The name was suggested by J. TICHÁ and M. TICHÝ.

MATTHIAS BRAUN wurde in Sauters, Tirol, geboren. Er war einer der bedeutendsten Bildhauer des böhmischen Barock und starb 54-jährig in Prag.

(6966) Vietoris
   E: 1991 Sep. 13 by L. D. SCHMADEL and F. BÖRNGEN at Tautenburg.
   B: L. D. SCHMADEL and H. HAUPT, MPC 32094, 1998 June 10.

Named in honor of LEOPOLD VIETORIS (b. 1891), professor emeritus of mathematics of the Leopold-Franzens-University in Innsbruck, on the occasion of his forthcoming 107th birthday. VIETORIS is the oldest full member of the Austrian Academy of Sciences and a highly decorated scientist. He made fundamental contributions to algebraic and set-theory topology, and he also wrote outstanding papers on the theory of real functions and on applied mathematics. Named by the first discoverer following a suggestion by H. HAUPT, who prepared the citation.

Professor VIETORIS, der noch bis ins hohe Alter ein begeisterter Schifahrer war, ist im Jahre 2002 als ältester Österreicher verstorben.
   R: NR Alm. d. ÖAW 152, 429 (2002).

(6973) Karajan
   E: 1992 Apr. 27 by S. UEDA and H. KANEDA at Kushiro.
   B: MPC 60728, 2007 Sep. 26.

HERBERT VON KARAJAN (1908–1989), born in Salzburg, was one of the best-known conductors of the twentieth century. He was music director of the Berlin Philharmonic Orchestra for thirty-five years and realized many famous recordings with that orchestra.

HERBERT RITTER VON KARAJAN war der Sohn des berühmten Hofbibliothekars und Universitätsprofessors für Deutsche Sprache und Literatur THEODOR GEORG VON KARAJAN, der auch Präsident der ÖAW war und selber aus einer griechischen Kaufmannsfamilie abstammte. In Österreich war HERBERT VON KARAJAN unter anderem Direktor der Wiener Staatsoper und vor allem in Zusammenhang mit den Salzburger Festspielen eine gut bekannte Persönlichkeit.

R: Biographien und Nachrufe, u. a. BdÖ 231.

(6999) **Meitner**

E: 1977 Oct. 16 by C. J. VAN HOUTEN and I. VAN HOUTEN-GROENEVELD on PALOMAR Schmidt plates taken by T. GEHRELS.

B: SBNC, MPC 27464, 1996 July 01.

Named in memory of LISE MEITNER (1878–1968), Austrian nuclear physicist. MEITNER was only the second woman to receive a doctorate from the University of Vienna, where she had been much inspired by BOLTZMANN. In 1912 she joined the Kaiser-Wilhelm-Institut in Berlin. Her collaboration with the director, OTTO HAHN, resulted in the discovery of protactinium, thereby demonstrating the existence of uranium-235. In 1938 MEITNER moved to Stockholm, where, with her nephew, OTTO FRISCH, she explained the presence of barium in the neutron-bombardment experiments of HAHN and STRASSMANN as due to fission, a term they coined. Named by the Small Bodies Names Committee.

LISE MEITNER wurde in Wien geboren und starb in Cambridge. Sie war kM(A) der ÖAW seit 1948.

R: NR Alm. d. ÖAW 119, 345 (1969); BdÖ 337.

(7074) **Muckea**

E: 1977 Sep. 10 by N. S. CHERNYKH at the Crimean Astrophysical Observatory.

B: V. K. ABALAKIN, MPC 38196, 2000 Jan. 24.

HERMANN MUCKE (b. 1935), director of the Urania Sternwarte and planetarium in Vienna, is well known for his work in ephemeris astronomy and the theory of astronomical phenomena. He has contributed much to the astronomical education of schoolchildren and adults. The name was suggested by V. K. ABALAKIN.

Professor HERMANN MUCKE ist bekannt für seine enorme kontinuierliche Tätigkeit in der himmelskundlichen Bildung, als Herausgeber der Zeitschrift „Der Sternenbote" und des Österreichischen Himmelskalenders seit 50 Jahren, Inhaber des von Professor THOMAS gegründeten „Astronomischen Büros" und Geschäftsführer des „Österreichischen Astronomischen Vereins". Schließlich ist er der

Leiter des unter seiner Ägide gebauten „Freilichtplanetariums Georgenberg" in Wien-Mauer, einer hervorragenden allgemein zugänglichen Beobachtungsstätte für freisichtige Himmelsbeobachtung. Bekannt ist auch die Reihe seiner Seminarpapiere und die Neuherausgabe von Canones der Sonnen- und Mondfinsternisse sowie der astronomischen Kurzkalender.

(7127) **Stifter**

E: 1991 Sep. 09 by F. BÖRNGEN and L. D. SCHMADEL at Tautenburg.

B: F. BÖRNGEN, MPC 29149, 1997 Feb. 22.

Named for the most famous Austrian narrator ADALBERT STIFTER (1805–1868). After formative years spent in the Bohemian Forest, he studied near the Benedictine Abbey in Kremsmünster, later living in Vienna and Linz. In his brilliant novels and epics (The Timber Forest, Rock Crystal, Indian Summer and Witiko) landscapes were described in a superb manner. STIFTER described the correlation of man and nature in a subtle manner, full of feeling. He dealt with questions of education, love and piety, and he was also engaged in painting and science. He gave full details of the total solar eclipse of 1842 July 8 as observed in Vienna.

R: Gedenkfeier Alm. d. ÖAW 118, 379 (1968).

(7146) **Konradin**

E: 1960 Sep. 24 by C. J. VAN HOUTEN and I. VAN HOUTEN-GROENEVELD on Palomar Schmidt plates taken by T. GEHRELS.

B: H. HAUPT and L. D. SCHMADEL, MPC 28622, 1996 Dec. 24.

Named in honor of KONRADIN FERRARI D'OCCHIEPPO (b. 1907), professor emeritus of astronomy of Vienna University and full member of the Austrian Academy of Sciences, on his forthcoming ninetieth birthday. Well known in Austria as a profound teacher and scientist, he is also a nobleman of extraordinary personal modesty and generosity. He has worked on variable stars, as well as on astrometric and calendrical problems and the history of astronomy. His main interest has been in the astronomical aspects of the Star of Bethlehem, a subject to which he devoted a book and many articles. Name suggested by H. F. HAUPT (who prepared the citation) and L. D. SCHMADEL.

Professor KONRADIN (Graf) FERRARI D'OCCHIEPPO, dessen Familienname schon für den Autoindustriellen FERRARI (KP 4122) gewidmet worden war, hat seine Ehrung durch die Benennung eines Planeten mit seinem Vornamen „Konradin" erhalten. Als Letzter seines Geschlechtes ist er 2007 im hundertsten Lebensjahr verstorben.

R: Zahlreiche Nachrufe, u. a. in Alm. d. ÖAW 157, 461 (2008). Der Sternenbote 50, 100 (2007) und eine ausführliche Biographie durch H. HAUPT in G. HEINDL, Wissenschaft und Forschung in Österreich, Verlag Lang, Frankfurt (2000), S 175–194.

(7400) **Lenau**

E: 1987 Aug. 21 by E. W. ELST at the European Southern Observatory.

B: MPC 59921, 2007 June 01.

NIKOLAUS LENAU (N. F. NIEMBSCH VON STREHLENAU, 1802–1850) was an Austrian poet. Unable to settle down to any profession, he started writing verses. Many poems (e.g., *Herbst*, *Schilflieder*) were inspired by his hopeless passion for SOPHIE VON LÖWENTHAL. LISZT used his *Faust* to compose the *Mephisto Walzer*.

LENAU war im Banat geboren und verstarb in Wien-Döbling.

R: diverse NR; Biographie BdÖ 285.

(7491) **Linzerag**

E: 1995 Sep. 23 at the Osservatorio San Vittore.

B: H. RAAB, MPC 30101, 1997 June 20.

Named on the occasion of the 50th anniversary of the Linzer Astronomische Gemeinschaft, a very active association of amateur astronomers in Austria. The number of this minor planet, written backwards, corresponds to the year when the association was founded. Name proposed by the discoverers following a suggestion from HERBERT RAAB, president of the association.

Die Linzer Astronomische Gemeinschaft ist ein sehr aktiver Verein in der oberösterreichischen Landeshauptstadt, siehe auch KP 3184 Raab.

(7583) **Rosegger**

E: 1991 Jan. 17 by F. BÖRNGEN at Tautenburg.

B: F. BÖRNGEN, MPC 30101, 1997 June 20.

Named for the Styrian author PETER ROSEGGER (1843–1918), in his lifetime extremely popular in Austria. Particularly known are his novels "*Die Schriften des Waldschulmeisters*" and "*Als ich noch der Waldbauernbub war*" describing the people and manners in the villages of his homeland.

R: Viele Nachrufe, siehe auch BdÖ 437.

(7624) **Gluck**

E: 1971 Mar. 25 by C. J. VAN HOUTEN and I. VAN HOUTEN-GROENEVELD on Palomar Schmidt plates taken by T. GEHRELS.

B: MPC 30802, 1997 Oct. 16.

Named for the German composer Ritter CHRISTOPH WILLIBALD VON GLUCK (1714–1787). After studying in Prague and Milan he traveled to London to meet HÄNDEL. He also worked with the traveling opera companies of MINGOTTI and LOCATELLI. In 1752 GLUCK became conductor in Sachsen-Hildburghausen and in Vienna. He composed 17 symphonies and 107 operas, including the three Italian reform operas Orfeo ed Euridice, Alceste and Paride ed Elena.

GLUCK gilt als einer der bedeutendsten Opern-Komponisten der zweiten Hälfte des 18. Jahrhunderts. Er lebte seit 1752 in Wien, wo er auch verstarb.

R: BdÖ 147.

(7722) **Firneis**
E: 1973 Sep. 29 by C. J. VAN HOUTEN and I. VAN HOUTEN-GROENEVELD on Palomar Schmidt plates taken by T. GEHRELS.

B: H. HAUPT, MPC 31298, 1998 Feb. 11.

Named in honor of MARIA GERTRUDE FIRNEIS (b. 1947), a professor of astronomy at the University of Vienna. She has worked on delicate astrometric and statistical problems and is an authority on the history of astronomy. She is also engaged in dating ancient events and correlating them with celestial phenomena. Name suggested and citation prepared by H. HAUPT.

Frau Professor FIRNEIS, in Wien geboren, ist heute eine vielseitige und sehr engagierte Forscherin und Lehrerin in den oben erwähnten Gebieten. Sie ist Mitglied einiger wissenschaftlicher Kommissionen, so auch der Kommission für Astronomie der ÖAW.

(7734) **Kaltenegger**
E: 1979 June 25 by E. F. HELIN and S. J. BUS at Siding Spring.

B: MPC 61764, 2008 Jan. 22.

Austrian-born LISA KALTENEGGER (b. 1977), at the Harvard-Smithsonian Center for Astrophysics since 2005, is involved in the use of space-based instruments to collect sufficient photons from extrasolar planets to characterize their physical and chemical composition and examine their potential habitability.

Frau Dr. KALTENEGGER wurde in Salzburg geboren. Sie studierte in Graz, zuerst an der Technischen Universität, und wurde dann 2004 an der Karl-Franzens-Universität Graz zur Dr. phil. *sub auspiciis praesidentis* promoviert. Verschiedentlich ausgezeichnet, arbeitet sie derzeit in den USA.

(7896) Švejk

E: 1995 Mar. 01 by Z. MORAVEC at Klet'.
B: MPC 34344, 1999 Apr. 02.

Named for a literary character created by JAROSLAV HAŠEK, Czech writer and humorist (1883–1923). His "good soldier" JOSEF ŠVEJK is devoted to the disintegrating Austrian-Hungarian monarchy regime, but by his simplicity (or perhaps his ingenuity) he helps to reveal its flaws.

JOSEF SCHWEJK (1891–1965) war ein tschechischer Tischler und die Hauptfigur der Geschichte von „Die Abenteuer des braven Soldaten Schwejk", oft gespielt von FRITZ MULIAR.

(7940) Erichmeyer

E: 1991 Mar. 13 at the Oak Ridge Observatory.
B: MPC 32095, 1998 June 10.

Named in honour of ERICH MEYER (b. 1951), Austrian amateur astronomer, on the occasion of his 20th anniversary as an astrometricist. Using a measuring engine he constructed himself, Meyer measured about 250 precise positions of minor planets and comets from photographic plates. Among the 2600 positions he derived after switching to CCD equipment in 1993 are some for the 1997 opposition of this object, thereby rendering it appropriate for numbering. An electrical engineer by profession, MEYER is also a well-known astrophotographer and populariser of astronomy.

E. MEYER ist ein bekannter oberösterreichischer Amateurastronom und KP-Beobachter. (Siehe auch KP 9097 Davidschlag.)

(8057) Hofmannsthal

E: 1971 Mar. 26 by C. J. VAN HOUTEN and I. VAN HOUTEN-GROENEVELD on Palomar Schmidt plates taken by T. GEHRELS.
B: MPC 34345, 1999 Apr. 02.

HUGO VON HOFMANNSTHAL (1874–1929) began writing poems at the age of 16. His lyrical and dramatic work reflects Austrian impressionism and symbolism. Together with RICHARD STRAUSS and MAX REINHARDT, he founded the Salzburger Festspiele. His best-known play is "Jedermann".

HUGO VON HOFMANNSTHAL, ein geborener Wiener, der auch in Rodaun gestorben und begraben ist, hat als freier Schriftsteller die Literaturszene des vorigen Jahrhunderts wesentlich beeinflusst.
R: BdÖ 200.

(8058) **Zuckmayer**

E: 1977 Oct. 16 by C. J. VAN HOUTEN and I. VAN HOUTEN-GROENEVELD on Palomar Schmidt plates taken by T. GEHRELS.

B: MPC 34345, 1999 Apr. 02.

Named for Jewish playwright CARL ZUCKMAYER (1896–1977). He wrote many comedies, his most famous play being "*Der Hauptmann von Köpenik*". As a young soldier he took part in World War I. His plays were later banned in Germany, so he emigrated to Austria and in 1938 moved via Switzerland to the U.S. During 1940–1946 he earned his living on a farm. There he wrote the much-discussed drama "*Des Teufels General*". Although he returned to Germany in 1947, he lived in Switzerland from 1958 onward.

ZUCKMAYER kam auch hier bei uns, durch seinen Aufenthalt in Österreich und seine Schriften, zu einem hohen Bekanntheitsgrad.

(9097) **Davidschlag**

E: 1996 Jan. 14 at Linz.

B: MPC 34628, 1999 May 04.

Named for a small rural village, some 10 km to the north of Linz, at the entrance to a region known as "Sterngartl", or "small garden of stars". This object is the first minor planet discovered at the amateur astronomical observatory that is located in this village.

Davidschlag hat sich in den letzten Jahren als Privatsternwarte voll etabliert und liefert beachtenswerte Resultate.

(9119) **Georgpeuerbach**

E: 1998 Feb. 18 at Linz.

B: MPC 34350, 1999 Apr. 02.

Named in memory of GEORG AUNPEKH VON PEUERBACH (1423–1461), professor at the University of Vienna, astronomer at the court of emperor FRIEDRICH III, mathematician, poet, early humanist and teacher of Regiomontanus. He discovered the magnetic declination, introduced sines into trigonometry and invented the foldable sundial, the first trustworthy pocket timepiece. Named on the occasion of the unveiling of a memorial tablet on St. Stephan's cathedral in Vienna, where Peuerbach is buried.

GEORG VON PEUERBACH, der selbst ein Schüler des JOHANNES VON GMUNDEN war, repräsentierte den ersten Höhepunkt der Astronomie in Österreich. Er spielte eine Rolle bei der Vorbereitung der Einführung des Gregorianischen Kalenders.

R: viele Biographien; Festschrift anlässlich der Gedenktafelauf-stellung; Der Sternenbote 4, 62 (1961); BdÖ 393; NDB 20, 281 (2001).

(9134) Encke

E: 1960 Sep. 24 by C. J. VAN HOUTEN and I. VAN HOUTEN-GROENEVELD on Palomar Schmidt plates taken by T. GEHRELS.

B: MPC 34350, 1999 Apr. 02.

Named in memory of JOHANN FRANZ ENCKE (1791–1865), eminent German astronomer. Beginning in 1816, ENCKE was assistant professor and director of the Seeberg Observatory, near Gotha. He computed the orbit of a short-period comet discovered by PONS and demonstrated that this object had been observed repeatedly. Later it was named comet 2P/Encke. In 1825 ENCKE was appointed director of the Berlin Observatory and member of the Berlin Academy of Sciences. ENCKE is most famous for editing the Berliner Astronomisches Jahrbuch from 1830 to 1866. This contained several very important papers on orbit determination and perturbation computations.

ENCKE war kM(A) der ÖAW ab 1848.

R: NR Alm. d. ÖAW 16, 245 (1866).

(9236) Obermair

E: 1997 Mar. 12 by E. MEYER at Linz.

B: MPC 34629, 1999 May 04.

Named in honor of the Austrian amateur astronomer ERWIN OBERMAIR (b. 1946), who, together with the discoverer, is co-owner of the private observatory in Davidschlag, near Linz. A technician by profession, OBERMAIR is also a well-known astro-photographer and popularizer of astronomy.

Siehe KP 7940 Erichmeyer und 9097 Davidschlag.

(9253) Oberth

E: 1971 Mar. 25 by C. J. VAN HOUTEN and I. VAN HOUTEN-GROENEVELD on Palomar Schmidt plates taken by T. GEHRELS.

B: SBNC, MPC 34351, 1999 Apr. 02.

Named in memory of the German pioneer of space flight HERMANN J. OBERTH (1894–1989). In his 1923 work entitled "Die Rakete zu den Planeträumen", OBERTH gave a thorough discussion of many phases of rocket travel, including the launching of payloads into earth orbit and the abnormal effects of pressure on the human body.

HERMANN OBERTH wurde 1894 in Hermannstadt (Siebenbürgen) ge-boren, studierte in Wien und entwickelte sich zu einem Raketenforscher und Raumfahrtpionier. 1945 wurde er nach Peenemünde verpflichtet und lebte dann bis 1989 in den USA.

R: BdÖ 376; Space Flight 32, 144 (1990).

(9307) **Regiomontanus**

E: 1987 Aug. 21 by F. BÖRNGEN at Tautenburg.

B: MPC 33795, 1999 Feb. 02.

Named for the German mathematician and astronomer JOHANNES REGIOMONTANUS (1436–1476), originally called J. MÜLLER, one of the most famous scholars of his time and trailblazer of the new world view. He improved mathematical methods and created modern trigonometry. He was sure that calculations of the orbits of celestial bodies could be improved essentially by new, more exact systematic observations. So in 1472 he founded the first German observatory in Nürnberg. His plan was interrupted by his early death. Calendars and ephemeris he calculated and published helped the sailors COLUMBUS, DA GAMA and VESPUCCI.

Der weltberühmte Astronom JOHANNES MÜLLER aus Königsberg in Franken, genannt REGIOMONTAN, hat in Wien studiert und war ein Schüler des GEORG VON PEUERBACH. Er kam auch nach Rom zum Zweck der Kalenderreform, konnte sie aber selbst nicht mehr erleben, weil er in Rom verstarb.

R: Zahlreiche Biographien, auch ADB 22, 564 (1885).

(9816) **von Matt**

E: 1960 Sep. 24 by C. J. VAN HOUTEN and I. VAN HOUTEN-GROENEVELD on Palomar Schmidt plates taken by T. GEHRELS.

B: H. HAUPT, MPC 43043, 2001 July 05.

ELISABETH BARONESS VON MATT (1762–1814), was the only female Austrian astronomer of her time with an international reputation. She worked mainly in positional astronomy. Her private observatory in Vienna was better equipped than the University Observatory. The name was suggested by H. HAUPT.

R: Über die Geschichte der BARONIN VON MATT ist wenig bekannt; ihre Erforschung ist derzeit noch im Gange. Vgl. W 17, 112.

(9824) **Marylea**

E: 1973 Sep. 30 by C. J. VAN HOUTEN and I. VAN HOUTEN-GROENEVELD on Palomar Schmidt plates taken by T. GEHRELS.

B: MPC 41570, 2000 Nov. 11.

MARY LEA SHANE, NÉE HEGER (1897–1983), wife of Lick Observatory director C. DONALD SHANE, was an astronomer in her own right. She identified the interstellar Na I absorption lines and discovered the first of the mysterious diffuse interstellar bands. She also established the Lick Observatory Archives.

Dr. MARY SHANE, geb. HEGER, war als Astronomin und bedeutende Gastgeberin auf der Licksternwarte während des Direktorates ihres Mannes Dr. C. DONALD SHANE und später als die Gründerin und Leiterin des „Lick Observatory Archive" eine gut bekannte Persönlichkeit. Sie sorgte auch freundlicherweise für den einen Verfasser dieser Arbeit (H. HAUPT) während seines dortigen Aufenthaltes als Fulbright-Stipendiat 1951/52. Erst nachher stellte sich heraus, dass die Familie HEGER aus Österreich stammte: Ihr Großvater war ein bedeutender Arzt am Allgemeinen Krankenhaus in Wien, der nach Amerika auswanderte und dort einen ranghohen Arztposten beim Militär bekleidete. Ihr Urgroßvater stammte aus Graz und konnte hier in den Matriken der Stadtpfarrkirche nachgewiesen werden.

R: NR Physics Today, Feb. 1984, p. 80; HLA XV (1984).

(9826) Ehrenfreund

E: 1977 Oct. 16 by C. J. VAN HOUTEN and I. VAN HOUTEN-GROENEVELD on Palomar Schmidt plates taken by T. GEHRELS.

B: C. J. VAN HOUTEN and I. VAN HOUTEN-GROENEVELD, MPC 41570, 2000 Nov. 11.

As both a biochemist and an astronomer, PASCALE EHRENFREUND (b. 1960), currently at the Leiden Observatory, qualified as an expert on several space missions investigating dust and organic molecules in space. As an APART fellow of the Austrian Academy of Sciences, she encouraged international cooperation.

(9833) Rilke

E: 1982 Feb. 21 by F. BÖRNGEN at Tautenburg.

B: MPC 34631, 1999 May 04.

Named for the Austrian poet RAINER MARIA RILKE (1875–1926), husband of the sculptor CLARA WESTHOFF. His monographs Auguste Rodin (1903) and Das Marienleben (1913), set to music by HINDEMITH, were written in Paris. Duineser Elegien was his greatest late work. In his lyric creations he always strove for sonorous language. Some of his work is strongly influenced by religious longings.

RILKE wurde in Prag geboren und verstarb nach ausgedehnten Reisen an Leukämie in Montreux in der Schweiz. Er hielt trotz seiner österreichisch-böhmischen Eltern Kärnten für das Land seiner Vorfahren. Sein lyrisches Werk erfuhr Auflagen in Millionenhöhe.

R: Zahlreiche Biographien, siehe auch BdÖ 431.

(9910) **Vogelweide**

E: 1973 Sep. 30 by C. J. VAN HOUTEN and I. VAN HOUTEN-GROENEVELD on Palomar Schmidt plates taken by T. GEHRELS.
B: MPC 34356, 1999 Apr. 02.
WALTHER VON DER VOGELWEIDE (c. 1170–1230) was probably born in Niederösterreich and died near Würzburg, Germany. A famous minstrel, his portrait appears in the Manesse-Handschrift from 1320. Quite a number of his songs with several verses have been preserved.

Der „Vogelweider" erhielt seine Ausbildung in Wien; er war der bedeutendste mittelhochdeutsche Lyriker und Sänger.
R: BdÖ 576 (Walther von der Vogelweide).

(10356) **Rudolfsteiner**

E: 1993 Sep. 15 by E. W. ELST at the European Southern Observatory.
B: MPC 39654, 2000 Mar. 20.
RUDOLF STEINER (1861–1925) was the editor of the scientific works of WOLFGANG GOETHE, and this inspired him to write his well-known work Die Philosophie der Freiheit (1894). In 1912 he founded the Anthroposophical Society upon the belief that there is a spiritual perception independent of the senses.

STEINER wurde in Kraljevic (Kroatien) geboren und ist in Dornach in der Schweiz gestorben. Er hat in seinem Leben mehrere Wandlungen durchgemacht, ist aber trotz mancher Widersprüche auch heute noch eine namhafte Autorität in der Anthroposophie. Seine wissenschaftlichen Äußerungen in Zusammenhang mit J. W. GOETHE (Farbenlehre) halten den neuen Ansichten nicht stand. Die „Waldorfschulen" gehen auf seine Gründung zurück und spielen heute eine Rolle in der „alternativen" (teils auch esoterischen) Erziehung.
R: BdÖ 518.

(10358) **Kirchhoff**

E: 1993 Oct. 09 by E. W. ELST at the European Southern Observatory.
B: MPC 39654, 1999 Nov. 23.
GUSTAV ROBERT KIRCHHOFF (1824–1887) was a German physicist who, together with ROBERT BUNSEN, founded the discipline of spectrum analysis. They demonstrated that an element gives off a characteristic colored light when heated to incandescence.

GUSTAV R. KIRCHHOFF war seit 1862 kM(A) der ÖAW.
  R: NR Alm. d. ÖAW 38, 193 (1888).

(10361) Bunsen

  E: 1994 Aug. 12 by E. W. ELST at the European Southern Observatory.
  B: MPC 39655, 1999 Nov. 23.

ROBERT WILHELM BUNSEN (1811–1899) was a German chemist who discovered the alkali-group metals cesium and rubidium. He also found an antidote to arsenic poisoning (1834) and invented the carbon-zinc electric cell (1841). He is best remembered by every chemistry student for the development of the Bunsen burner.

ROBERT W. BUNSEN war kM(A) der ÖAW ab 1848, ab 1899 EM.
  R: NR Alm. d. ÖAW 50, 286 (1900).

(10634) Pepibican

  E: 1998 Apr. 08 by L. ŠAROUNOVÁ at Ondřejov.
  B: MPC 36129, 1999 Sep. 28.

JOSEF ("PEPI") BICAN (b. 1913), proclaimed the century's best center-forward by the International Federation of Soccer Historians and Statisticians, represented Austria in 19 and Czechoslovakia in 14 international matches and scored more than 5000 goals in his career. After retirement he developed an interest in astronomy.

JOSEF „PEPI" BICAN ist in Wien geboren und 2001 in Prag verstorben. Er war einer der ganz großen Fußballspieler Europas, der sowohl in der österreichischen wie der tschechoslowakischen Nationalmannschaft spielte und mit diesen viele internationale Erfolge erzielte.

(10709) Ottofranz

  E: 1982 Jan. 24 by E. BOWELL at the Lowell Observatory's Anderson Mesa Station.
  B: L. H. WASSERMAN, MPC 53469, 2005 Jan. 25.

Lowell Observatory astronomer OTTO G. FRANZ (b. 1931) has studied binary stars using photography, area scanning techniques, speckle interferometry, by means of spectroscopic measurements of radial velocity, and using the Fine Guidance Sensors of the Hubble Space Telescope. The name was suggested by L. H. WASSERMAN.

OTTO FRANZ ist in Österreich wohlbekannt als guter Freund; er studierte in Wien und wurde 1955 *sub auspiciis praesidentis* promoviert. Er ging dann in die USA, wo er durch K. A. STRAND sehr gefördert wurde. Nach mehreren Zwischenstationen arbeitete er schließlich viele Jahre lang am Lowell Observatory in Flagstaff, Arizona, wo er auch über die aktive Dienstzeit hinaus seine erfolgreiche Tätigkeit weiter fortsetzt.

(10763) **Hlawka**

E: 1990 Oct. 12 by L. D. SCHMADEL and F. BÖRNGEN at Tautenburg.
B: L. D. SCHMADEL, MPC 42360, 2001 Mar. 09.

EDMUND HLAWKA (b. 1916) is the most famous living Austrian mathematician and a prominent university teacher in Vienna. He contributed important theorems to the geometry of numbers, to the theory of uniform distribution, and to numerical integration. The citation was prepared by H. HAUPT.

Professor HLAWKA ist ein angesehenes Mitglied der ÖAW (seit 1956 k.M., seit 1959 w.M.) und war stets ein wohlwollender Freund der Astronomie, die er in verschiedenen Funktionen an der Universität Wien und an der Akademie nachhaltig unterstützte.

(10782) **Hittmair**

E: 1991 Sep. 12 by L. D. SCHMADEL and F. BÖRNGEN at Tautenburg.
B: L. D. SCHMADEL and H. HAUPT, MPC 42360, 2001 Mar. 9.

OTTO HITTMAIR (b. 1924) is a well-known Austrian theoretical physicist. He was president of the Austrian Academy of Sciences in Vienna. His main contributions were on problems of quantum theory, superconductivity and general field theory. The name was suggested by L. D. SCHMADEL, and the citation was prepared by H. HAUPT.

Professor HITTMAIR, der 1977/79 auch Rektor der Technischen Universität Wien war, hat sich in der Akademie stets für die Belange der Kommission für Astronomie (KfA) eingesetzt. Seit 1970 war er w.M. Nach seiner Emeritierung ging er wieder in seine Tiroler Heimat, wo er im Herbst 2003 bei einer Bergtour tödlich verunglückte.

R: NR Alm. d. ÖAW 153, 405 (2003).

(10787) **Ottoburkard**

E: 1991 Oct. 04 by L. D. SCHMADEL and F. BÖRNGEN at Tautenburg.
B: L. D. SCHMADEL, MPC 42361, 2001 Mar. 09.

OTTO M. BURKARD (b. 1908), professor emeritus of meteorology and geophysics of the University of Graz, was one of the founders of the Space Research Institute of the Austrian Academy of Sciences. The name was suggested by L. D. SCHMADEL, and the citation was prepared by H. HAUPT.

Professor BURKARD war im Jahre 1970 Rektor der Karl-Franzens-Universität Graz und hat sich als solcher und darüber hinaus als w.M. der ÖAW ständig für die Astronomie in Graz und ihre Forschungen, auch auf dem Sonnenobservatorium Kanzelhöhe in Kärnten, eingesetzt. Seine eigenen Forschungen betrafen vor allem die Ionosphäre, für die er eine eigene Beobachtungsstation einrichtete, die bis zu seiner Emeritierung in Betrieb war. Seit 1962 war er k.M. und seit 1969 ist er w.M. der ÖAW.

(10957) **Alps**
E: 1960 Sep. 24 by C. J. VAN HOUTEN and I. VAN HOUTEN-GROENEVELD on Palomar Schmidt plates taken by T. GEHRELS.
B: MPC 48392, 2003 May 01.
The Alps form a great mountain chain stretching from the Mediterranean Sea between southern France and Italy through Switzerland to eastern Austria.

(11019) **Hansrott**
E: 1984 Apr. 25 by A. MRKOS at Klet'.
B: J. MEEUS, MPC 53174, 2004 Nov. 26.
Much admired by GUSTAV MAHLER, the Austrian composer HANS ROTT (1858–1884) completed his Symphony in E Major in 1880. ROTT died at the age of 26 in a Viennese lunatic asylum. The name was suggested by J. MEEUS.

Das tragische Schicksal dieses in jungen Jahren verstorbenen Komponisten, dessen Werke erst nach seinem Tode veröffentlicht worden sind, wurde hier vom belgischen Astronomen J. MEEUS gewürdigt, der den Namen für diesen KP vorgeschlagen hat.

(11299) **Annafreud**
E: 1992 Sep. 22 by E. W. ELST at the European Southern Observatory.
B: MPC 39657, 2000 Mar. 20.
ANNA FREUD (1895–1982), the youngest daughter of SIGMUND FREUD, escaped with her father in 1938 Austria and settled in London. In 1936 she published *"Das Ich und die Abwehr-Mechanismen"*. She is considered the founder of child psycho-analysis.
R: BdÖ 123.

(11338) **Schiele**
E: 1996 Oct. 13 by J. TICHÁ and M. TICHÝ at Klet'.
B: MPC 41031, 2000 July 26.

EGON SCHIELE (1890–1918), Austrian expressionist painter noted for the eroticism of his gurative works, lived and worked in Ceskỳ Krumlov from 1907 to 1917.

SCHIELE wurde in Tulln geboren, er lebte und arbeitete vorwiegend in Wien und Niederösterreich, wo er sich als „Spitze der expressionistischen Avantgarde" profilierte. Er starb 28-jährig an der Spanischen Grippe.

R: Viele Nachrufe, u. a. BdÖ 459.

(11519) **Adler**
E: 1991 Apr. 08 by E. W. ELST at the European Southern Observatory.

B: MPC 40707, 2000 May 23.

ALFRED ADLER (1870–1937) was an Austrian psychiatrist who introduced the term "inferiority feeling" (inferiority complex) into psychology: mental health is characterized by reason and social interest, mental disorder by inferiority, self-concern and superiority over others. In 1934 the Austrian government closed his 30 clinics.

ALFRED ADLER wurde in Wien geboren und ist in Schottland gestorben. Nach der Trennung von seinem Lehrer SIGMUND FREUD (1912) vertrat er weiter seine tiefenpsychologische Schule als Professor in Wien und New York. Sein Schüler und zeitweiliger Mitarbeiter MANES SPERBER lieferte mehrere Schriften über sein Leben und seine Werke.

R: BdÖ 7/8.

(11524) **Pleyel**
E: 1991 Aug. 02 by E. W. ELST at the European Southern Observatory.

B: MPC 41031, 2000 July 26.

IGNAZ JOZEF PLEYEL (1757–1831) was an Austro-French composer who became famous as a piano builder. In 1795 he settled in Paris, where he founded a piano-manufacturing company in 1807. The famous Polish-French composer CHOPIN preferred the Pleyel grand piano for performing his compositions.

I. J. PLEYEL wurde in Ruppersthal in Niederösterreich geboren und starb auf einem Landgut in der Nähe von Paris. Er war Kompositionsschüler von J. HAYDN und wirkte als Kapellmeister in verschiedenen Positionen in Österreich, Italien, Frankreich und England, wo er zahlreiche Werke, darunter zwei Opern, komponierte.

R: BdÖ 401.

(11573) **Helmholtz**

E: 1993 Sep. 20 by F. BÖRNGEN and L. D. SCHMADEL at Tautenburg.

B: MPC 39658, 2000 Feb. 22.

HERMANN LUDWIG FERDINAND VON HELMHOLTZ (1821–1894), a German doctor, physiologist and physicist, was one of the most famous naturalists of the nineteenth century. In 1847 he extended ROBERT MAYER's law of energy conservation to all known natural phenomena.

H. HELMHOLTZ war seit 1860 kM(A) und seit 1872 EM der ÖAW.

R: NR Alm. d. ÖAW 45, 284 (1895).

(11583) **Breuer**

E: 1994 Aug. 12 by E. W. ELST at the European Southern Observatory.

B: MPC 40707, 2000 May 23.

JOSEF BREUER (1842–1925) was an Austrian physician who anticipated the process of psychoanalysis. In the famous case study of "Anna O.", BREUER concluded that neurotic symptoms result from unconscious processes.

JOSEF BREUER lebte und starb in Wien. Er arbeitete schon vor und dann mit SIGMUND FREUD über die Hysterie. Er war k.M. der ÖAW.

R: NR Alm. d. ÖAW 78, 216 (1928), u. a. BdÖ 48.

(11757) **Salpeter**

E: 1960 Sep. 24 by C. J. VAN HOUTEN and I. VAN HOUTEN-GROENEVELD on Palomar Schmidt plates taken by T. GEHRELS.

B: MPC 51187, 2004 Mar. 06.

EDWIN E. SALPETER (b. 1924), born in Austria, educated in Australia and the U.K., has been at Cornell University since 1948. A generalist, he worked with HANS BETHE in quantum electro-dynamics and nuclear physics and is best known for the triple-α helium-burning reaction and for the initial mass function of stars.

Der aus Österreich stammende bedeutende Astrophysiker EDWIN E. SALPETER emigrierte mit seiner Familie nach Australien, studierte in Sydney und später in Birmingham (England) und war Professor an der Cornell University in den USA.

(12002) **Suess**

E: 1996 Mar. 19 by P. PRAVEC and L. ŠAROUNOVÁ at Ondřejov.

B: H. RAAB, MPC 38201, 2000 Jan. 24.

The Austrian geologist FRANZ EDUARD SUESS (1867–1941), professor at the Technical College in Prague and the University of Vienna, worked on crystalline bedrock of the Bohemian Mass and made fundamental studies of the moldavites. He introduced the word "tektite". The name was suggested by H. RAAB.

FRANZ EDUARD SUESS wurde in Wien geboren und starb ebenda. Er war der Sohn von EDUARD SUESS (1813–1908) und als Geologe von 1908–1910 Professor an der Technischen Hochschule in Prag, von 1911–1938 Professor an der Universität Wien. Er war k.M. seit 1911 und ab 1915 w.M. der ÖAW.
> R: NR Alm. d. ÖAW 95, 319 (1945).

(12244) **Werfel**
> E: 1988 Sep. 08 by F. BÖRNGEN at Tautenburg.
> B: MPC 40708, 2000 May 23.

The Austrian author FRANZ WERFEL (1890–1945) was a playwright who achieved international success as a novelist and essayist (*Die vierzig Tage des Musa Dagh; Das Lied von Bernadette; Between heaven and earth*). After his rejection by the Prussian Academy of Poetry, he fled to the U.S. in 1940.

Der in Prag geborene FRANZ WERFEL kam nach dem Ersten Weltkrieg nach Wien und wirkte hier als freier Schriftsteller. 1926 erhielt er den Grillparzer-Preis der ÖAW für sein dramatisches Werk „Juarez und Maximilian" (Alm. d. ÖAW 76, 123). Er heiratete die Witwe GUSTAV MAHLERS und emigrierte 1938 nach Frankreich und später in die USA, wo er 1945 in Beverly Hills verstarb. 1945 wurde er in einem Ehrengrab am Wiener Zentralfriedhof beigesetzt. Er hinterließ zahlreiche bedeutende Werke.
> R: Viele Nachrufe, BdÖ 586.

(12320) **Loschmidt**
> E: 1992 Aug. 08 by E. W. ELST at Caussols.
> B: MPC 51188, 2004 June 14.

The Czech physicist JOSEF LOSCHMIDT (1821–1895) used the kinetic theory of gases to get the first reasonable estimate of molecular size. To distinguish it from Avogadro's number, the term Loschmidt's number has been reserved for the number of molecules in one cubic centimetre of a gas under standard conditions.

LOSCHMIDT war in Böhmen geboren worden und lebte als Physiker und Chemiker in Wien, wo er von 1871 bis 1892 ord. Professor an der Universität war. Er starb 1895 in Wien. Er war k.M. seit 1867, w.M. der ÖAW seit 1870.
> R: NR Alm. d. ÖAW 46, 258 (1896), BdÖ 305.

(12568) **Kuffner**

E: 1998 Nov. 11 by K. KORLEVIĆ at Višnjan.
B: MPC 56613, 2006 Apr. 13.

MORIZ VON KUFFNER (1854–1939) was a brewer, alpinist and the founder of a private observatory in Vienna. The Kuffner Observatory was a leading scientific institution in the late nineteenth century and is now used for public education. It is serving as host for the May 2006 Meeting on Asteroids and Comets in Europe.

Der in Wien geborene Edle MORIZ VON KUFFNER erbte eine Brauerei in Ottakring und ein Millionenvermögen von seinem kinderlos verstorbenen Onkel. Das erlaubte ihm, seinen Liebhabereien, darunter besonders der Astronomie, nachzugehen. Die von ihm 1884–86 gebaute Sternwarte oberhalb des Ottakringer Friedhofes, die bestens ausgestattet war, erlaubte ihm die Anstellung tüchtiger junger Männer, die später berühmte Astronomen wurden (wie z. B. SCHWARZSCHILD) und die Herausgabe von Druckwerken. Zeitweise war diese Sternwarte von größerer Bedeutung als die Wiener (Universitäts-)Sternwarte. KUFFNER starb 1939 unter traurigen persönlichen Verhältnissen in Zürich. Nach komplizierten Vorgängen während der Nachkriegszeit (vgl. W. W. WEISS) entstand schließlich die Kuffner'sche Volkssternwarte, die heute eine wichtige Rolle in der Wiener Volksbildung spielt.

R: ÖBL 4, 330 (1969); BdÖ 268; W. W. Weiss.

(12799) **von Suttner**

E: 1995 Nov. 26 at Klet'.
B: MPC 43762, 2001 Nov. 01.

BERTHA FÉLICIE SOPHIE, Freifrau VON SUTTNER, née countess KINSKY (1843–1914), was an Austrian novelist and one of the first notable woman pacifists. She is credited with influencing ALFRED NOBEL in the establishment of the Nobel Prize for Peace, of which she was the recipient in 1905.

Dem Bekanntheitsgrad von BERTA VON SUTTNER und ihrer Bedeutung als Friedensstifterin ist nichts hinzuzufügen. Geboren in Prag, starb sie in Wien wenige Wochen vor Ausbruch des Ersten Weltkrieges.

R: Zahlreiche Nachrufe, BdÖ 533.

(13092) **Schrödinger**

E: 1992 Sep. 24 by F. BÖRNGEN and L. D. SCHMADEL at Tautenburg.
B: MPC 40710, 2000 May 23.

The Austrian physicist ERWIN SCHRÖDINGER (1887–1961), born in Vienna, founded wave mechanics in 1926. Later he worked in relativistic quantum mechanics, the theory of gravity and unified field theory. Together with DIRAC, he received the 1933 Nobel prize in physics.

ERWIN SCHRÖDINGER war ab 1928 k.M. und ab 1956 w.M. der ÖAW. Er erhielt als Erster den höchsten nach ihm benannten Preis der mathematisch-naturwissenschaftlichen Klasse der ÖAW. Dieser wird seither regelmäßig verliehen.

R: Zahlreiche Nachrufe, bes. Alm. d. ÖAW 111, 402 (1961); u. a. BdÖ 482.

(13093) Wolfgangpauli
E: 1992 Sep. 21 by F. BÖRNGEN at Tautenburg.
B: MPC 40710, 2000 May 23.

The Austrian physicist WOLFGANG PAULI (1900–1958), born in Vienna, was co-founder of quantum theory. He discovered the Pauli principle, which explains the level structure of atoms. He received the 1945 Nobel prize in physics.

Als Sohn eines berühmten Vaters, des Mediziners WOLFGANG PAULI, kam PAULI bereits 1922 als Assistent nach Göttingen, arbeitete dann in Kopenhagen und Hamburg, bevor er Professor an der ETH Zürich wurde, wo er (nach mehreren zwischenzeitlichen Aufenthalten in den USA) auch verstarb. Er war auch der Entdecker des Neutrinos.

R: BdÖ 388.

(13122) Drava
E: 1994 Feb. 07 by E. W. ELST at the European Southern Observatory.
B: MPC 40710, 2000 May 23.

The Drava is a major tributary of the Danube river. It flows from its source in the Italian Alps through the Austrian federal states of Tirol and Kärnten. Huns and Slavs invaded the Alpine countries through its valley.

Die Drau (so der deutsche Name) ist mit ihren markanten Abschnitten eine besondere landschaftliche Schönheit in Österreich und durch die an ihr befindlichen Kraftwerke ein bedeutender wirtschaftlicher Faktor des Landes.

(13531) Weizsäcker
E: 1991 Sep. 13 by F. BÖRNGEN and L. D. SCHMADEL at Tautenburg.
B: MPC 40711, 2000 May 23.

CARL FRIEDRICH FREIHERR VON WEIZSÄCKER (b. 1912), German physicist and philosopher, was involved in nuclear physics, quantum theory and astronomy. Together with H. BETHE, he explained the radiation energy of stars by processes of nuclear fusion. He also developed a theory for the formation of the solar system.

CARL FRIEDRICH VON WEIZSÄCKER (1912–2007) war kM(A) ab 1966. Er war Ehrendoktor zahlreicher Universitäten und Mitglied vieler Akademien.

R: Alm. d. ÖAW 157, 133 (2008); NR später.

(13682) Pressberger
E: 1997 Aug. 10 by E. MEYER and H. RAAB at Linz.
B: MPC 41033, 2000 July 26.

RUDOLF PRESSBERGER (b. 1942) invented an improved telescope fork mount, named the "Austria Mount". First released in 1986, it features a right-ascension axis built inside the fork, works without ball bearings and uses friction drives. Using this design, PRESSBERGER also built a 1.0-m Ritchey-Chrétien telescope himself.

Die Pressbergerschen Teleskopmontierungen sind in Österreich weit verbreitet und hoch angesehen. Seine eigene Sternwarte in Klosterneuburg-Kierling, die „Purgathofer-Sternwarte" (siehe KP 5341) wird nach seinem frühen Tod in den Bergen im August 2001 für qualifizierte Beobachtungen weitergeführt.

R: Der Sternenbote 44, 177 (2001), JB der Univ.-Sternwarte Wien.

(13977) Frisch
E: 1992 Apr. 29 by F. BÖRNGEN at Tautenburg.
B: MPC 41033, 2000 July 26.

The Austrian zoologist KARL RITTER VON FRISCH (1886–1982) did fundamental investigations on fish and honey-bees. He studied the orientation of bees following colors, shapes, odors, the sun and polarized light. He received the 1973 Nobel prize for medicine.

Der Zoologe KARL VON FRISCH war gebürtiger Wiener. Er war seit 1919 Professor und wirkte von 1946–1950 auch in Graz. Er wurde 1939 k.M., dann w.M. und ab 1954 EM der ÖAW.

R: NR Alm. d. ÖAW 133, 297 (1983); BdÖ 130.

(14057) Manfredstoll
E: 1996 Jan. 15 by E. MEYER and E. OBERMAIR at Linz.
B: MPC 41034, 2000 July 26.

The Viennese computer specialist MANFRED STOLL (b. 1938) is an expert in the practical application of computers in astronomy. Among other projects, he worked on software for digitizing photographic plates. He also developed a modern telescope drive system and control software, which is now also used at the observatory in Linz.

STOLL promovierte in Wien bei PURGATHOFER (KP 5341), war dann ab 1977 Assistent und ab 1983 Oberrat an der Wiener Universitäts-Sternwarte, dann bis zu seiner Pensionierung Ministerialrat im Wissenschaftsministerium.

R: STOLL, MANFRED: Das 1m-RC-Teleskop der Purgathofer-Sternwarte: Seine Mechanik, seine Optik und seine Einsatzmöglichkeiten, Der Sternbote 2005, 5.

(14877) Zauberflöte

E: 1990 Nov. 19 by E. W. ELST at the European Southern Observatory.

B: MPC 56613, 2006 Apr. 13.

MOZART's opera *Die Zauberflöte* ("The Magic Flute") celebrates love, forgiveness and the brotherhood of men. It was his last opera, composed in 1791. The opera depicts many symbols of Freemasonry, since MOZART was himself a Freemason. This naming is on the occasion of the 250th anniversary of MOZART's birth.

(14977) Bressler

E: 1997 Sep. 26 by E. MEYER at the Davidschlag Observatory at Linz.

B: MPC 41387, 2000 Oct. 13.

The Austrian amateur astronomer MARTIN BRESSLER (b. 1912) started his astrometric program on minor planets in 1982. Always eager to learn new techniques, he enthusiastically switched from photographic emulsions to a CCD in 1993.

(15318) Innsbruck

E: 1993 May 24 by C. S. SHOEMAKER at Palomar.

B: P. TEUTSCH and R. WEINBERGER, MPC 42363, 2001 Mar. 09.

Innsbruck, capital of Tyrol in the heart of the Alps on the great route from Italy, has a history deeply connected with the Habsburg dynasty. This city in western Austria lays claim to the 1936 Nobel prize winner V. F. HESS and hosted two winter Olympic games. The name was suggested by P. TEUTSCH and R. WEINBERGER.

Hier ist ein gebührender Tribut gezollt für die Tiroler Landeshauptstadt (vgl. KP 6439) mit ihren Einwohnern, den kunsthistorischen Schönheiten und ihrer großen Bedeutung in der Vergangenheit und Gegenwart.

(15378) **Artin**

E: 1997 Aug. 07 by P. G. COMBA at Prescott.
B: MPC 41573, 2000 Nov. 11.

EMIL ARTIN (1898–1962) was an Austrian-German mathematician who lived for some years in the United States and made important contributions to abstract algebra, including the theories of rings and of braids.

EMIL ARTIN, geboren in Wien, gestorben in Hamburg, war ein bedeutender österreichischer Mathematiker, der seiner jüdischen Frau wegen 1937 in die USA emigrierte.

(15949) **Rhaeticus**

E: 1998 Jan. 17 by E. MEYER and E. OBERMAIR at the Davidschlag Observatory, Linz.
B: MPC 42363, 2001 Mar. 09.

RHAETICUS (GEORG JOACHIM VON LAUCHEN, 1514–1576) was a humanist, physician, mathematician and astronomer at the universities of Vienna, Leipzig and Wittenberg. He summarized and popularized the work of his teacher COPERNICUS, initiating the first printing of *De revolutionibus orbium coelestium.*

RHAETICUS wurde in Feldkirch in Vorarlberg geboren und ist in Kaschau (heute Košice) in der Slowakei im Jahre 1576 gestorben. Außer seinen oben angeführten Werken erstellte er auch eine zehnstellige Tafel der goniometrischen Funktionen und leistete Vorarbeiten für die Erfindung der Logarithmentafeln.
R: BdÖ 428.

(15955) **Johannesgmunden**

E: 1998 Jan. 26 by E. MEYER at the Davidschlag Observatory, Linz.
B: MPC 42363, 2001 Mar. 09.

JOHANNES VON GMUNDEN (1380/1384–1442), a priest, humanist, mathematician and astronomer at the University of Vienna, brought the mathematical and astronomical knowledge of Arabia to Europe. Known for his precise ephemerides, he also published the first printed calendar in German.

JOHANNES VON GMUNDEN lebte in der ersten Blütezeit der Wiener Astronomie wie GEORG VON PEUERBACH (KP 9119) und REGIOMONTAN (KP 9307), die die „Wiener Schule" bildeten und in einem Schüler-Lehrer-Verhältnis standen.
R: Sitzungsber. der phil.-hist. Klasse ÖAW 222, 1–93 (1943), auch BdÖ 222.

(15963) **Koeberl**
E: 1998 Feb. 06 by E. W. ELST at the European Southern Observatory.
B: MPC 56959, 2006 June 13.
CHRISTIAN KOEBERL (b. 1959) is a professor of geochemistry and cosmochemistry at the University of Vienna. His work involves the study of impact structures due to asteroidal bodies, as well as the investigation of the chemistry of tektites, impact glasses and lunar meteorites.

CHRISTIAN KOEBERL wurde in Wien geboren und hat dort studiert, promovierte aber in Graz. Er ist sehr aktiv und publiziert fleißig. Seit 2004 ist er k.M. und seit 2006 w.M. der ÖAW.

(16355) **Buber**
E: 1975 Oct. 29 by F. BÖRNGEN at Tautenburg.
B: MPC 41573, 2000 Nov. 11.
MARTIN BUBER (1878–1965), Austrian-born Jewish philosopher and author, was a teacher of religious science, ethics and social philosophy. From 1938 he taught in Jerusalem, where he stood up for the peaceful coexistence of Arabs and Jews. His Hebrew-German version of the Bible shows a unique diction and exegesis.
R: Kurze Biogr. in BdÖ 52.

(16398) **Hummel**
E: 1982 Sep. 24 by F. BÖRNGEN at Tautenburg.
B: MPC 42364, 2001 Mar. 09.
JOHANN NEPOMUK HUMMEL (1778–1837), famous Austrian pianist and versatile composer, MOZART's pupil and BEETHOVEN's friend, made numerous concert tours. He was appointed conductor of the court orchestra in Weimar in 1819 and held this position until his death. His tomb is in the historical churchyard in Weimar.

HUMMEL wurde in Pressburg geboren und starb in Jerusalem.
R: Kurze Biogr. in BdÖ 212.

(16445) **Klimt**
E: 1989 Apr. 03 by E. W. ELST at the European Southern Observatory.
B: MPC 58595, 2007 Jan. 06.
The Austrian painter GUSTAV KLIMT (1862–1918) was one of the most prominent members of the *Vienna Art Nouveau* movement. His paintings, characterized by elegant gold and colorful ornamentation, express subtle erotic feelings, as in *Die Jungfrau* (1907) and *Dana* (1913).

Der Wiener GUSTAV KLIMT gilt als Hauptvertreter des Jugendstils. Er schuf zahlreiche Bilder für die Sezession, das Kunsthistorische Museum, das Burgtheater und die Universität Wien, die alle zu den sehenswertesten Kunstwerken Österreichs gehören. Er starb 1918 in Wien.

(16705) **Reinhardt**

E: 1995 Mar. 04 by F. BÖRNGEN at Tautenburg.
B: MPC 42364, 2001 Mar. 09.

The Austrian stage director and theatre manager MAX REINHARDT (MAX GOLDMANN, 1873–1943) worked mainly in Berlin and Vienna. He was a cofounder of the "Salzburger Festspiele". His productions of classic dramas caused an enormous stir. In 1933 he emigrated from Germany.

Der in Baden bei Wien geborene Schauspieler hat in Österreich und bei vielen Auslandsaufenthalten der Bühnenkunst neue Wege gewiesen, die in zahlreichen Biographien und Nachrufen dokumentiert sind. Er war u. a. auch Gründer des „Reinhardt-Seminars". Er war in zweiter Ehe mit HELENE THIMIG verheiratet und starb 1943 in New York.

R: u. a. BdÖ 422.

(16802) **Rainer**

E: 1997 Sep. 25 by E. MEYER at the Davidschlag Observatory, Linz.
B: MPC 45234, 2002 Mar. 28.

Suffering from a serious heart disease since his birth, RAINER GEBETSROITHER (1976–1998) devoted his life to observations of nature as well as to the history and technology of railways. His parents KARIN and UWE are long-term members of the Linzer Astronomische Gemeinschaft.

Eine ehrende Erinnerung an den 22-jährigen Verstorbenen und seine Eltern.

(17459) **Andreashofer**

E: 1990 Oct. 13 by F. BÖRNGEN and L. D. SCHMADEL at Tautenburg.
B: MPC 42365, 2001 Mar. 09.

Innkeeper ANDREAS HOFER (1767–1810) headed the Tyrolese popular rising against French occupation and was executed by a firing squad on order of NAPOLEON. His patriotic and heroic engagement is the subject of numerous dramatic plays, stories and poems, notably by ROSEGGER, EICHENDORFF and KOERNER.

Die tragische Figur des Tiroler Freiheitshelden, als „Sandwirt" in Sankt Leonhard in Passeier (Südtirol) geboren und in Mantua durch Erschießen hingerichtet, ist wohl allen in Österreich bekannt. Den zahlreichen in der Citation erwähnten Geschichten ist nichts hinzuzufügen.

R: BdÖ 198.

(17460) **Mang**

E: 1990 Oct. 10 by L. D. SCHMADEL and F. BÖRNGEN at Tautenburg.

B: L. D. SCHMADEL, MPC 52324, 2004 July 13.

HERBERT MANG (b. 1942) is professor of material sciences at the Vienna University of Technology. He is a well-known expert in computational mechanics and a friend of astronomy. MANG currently serves as president of the Austrian Academy of Sciences in Vienna. The name was suggested by the first discoverer.

Nach Beendigung seiner Amtsperiode als Präsident der ÖAW im Jahre 2006 ist Professor MANG nach wie vor intensiv in die Arbeit der Akademie und seines Institutes eingebunden, wie vor allem seine überreiche Publikationsliste zeigt.

(17488) **Mantl**

E: 1991 Oct. 02 by L. D. SCHMADEL and F. BÖRNGEN at Tautenburg.

B: L. D. SCHMADEL, MPC 52324, 2004 July 13.

WOLFGANG MANTL (b. 1939) is professor of jurisprudence and constitutional law at the University of Graz. MANTL is a leading expert in building bridges to the neighboring countries of southeast Europe. He is chairman of the Austrian Board of Science and a lover of astronomy. The name was suggested by the first discoverer.

Professor MANTL ist mit der Erreichung des Emeritierungsalters im Jahre 2007 hoch geehrt worden. Er übt aber weiterhin zahlreiche wissenschaftliche Tätigkeiten und hohe Funktionen aus. Seit 1999 ist er w.M. der ÖAW.

(17489) **Trenker**

E: 1991 Oct. 02 by F. BÖRNGEN and L. D. SCHMADEL at Tautenburg.

B: MPC 42365, 2001 Mar. 09.

LUIS TRENKER (1892–1990), initially a herdsman, mountain guide and ski instructor in his South Tyrolese alpine homeland, later had great success as a scriptwriter, film director, screen actor and author. During 1928–1940 he lived in Berlin. His novels, which described and praised his native country, later became very popular.

Die bergsteigerischen Rekorde von LUIS TRENKER sind auch im heutigen Österreich noch lebendig.

R: BdÖ 553.

(17597) **Stefanzweig**

E: 1995 Mar. 04 by F. BÖRNGEN at Tautenburg.

B: MPC 42676, 2001 May 09.

The Austrian STEFAN ZWEIG (1881–1941), biographer, essayist and writer, communicated with world figures and had great confidence in the good strengths of humanity. His books were translated into more than 20 languages. Having emigrated in 1934, he suffered from being abroad and eventually committed suicide.

Der in Wien geborene Schriftsteller studierte und promovierte auch in Wien. Als Kriegsgegner bereiste er mehrere Kontinente und lebte dann vorwiegend in Salzburg. 1938 wanderte er nach Amerika aus. Gemeinsam mit seiner zweiten Frau HILDE schied er bei Rio de Janeiro aus dem Leben.

R: Zahlreiche Biographien und Nachrufe; auch BdÖ 615.

(18032) **Geiss**

E: 1999 June 20 by the Lowell Observatory Near-Earth Object Search at the Anderson Mesa Station.

B: T. C. Owen, MPC 56960, 2006 Jun. 13.

JOHANNES GEISS (b. 1926) is a leading Swiss space scientist. He was PI on the Solar Wind Composition experiment aboard Apollo 11–16. The recipient of medals from learned societies of the U.S. and Europe, GEISS, is a foreign member of the U.S. Academy of Sciences. The name was suggested by T. C. OWEN.

Prof. Dr. Dr. h.c. J. GEISS, geboren in Stolp (Nordpommern), ist auch kM(A) der ÖAW seit 1989 und war in Österreich als Evaluator tätig.

(18395) **Schmiedmayer**

E: 1992 Sep. 21 by L. D. SCHMADEL and F. BÖRNGEN at Tautenburg.

B: L. D. SCHMADEL, MPC 53176, 2004 Nov. 26.

JÖRG SCHMIEDMAYER (b. 1960) is an Austrian physicist and a leading expert in the field of quantum optics. A professor at the Heidelberg University, he is also an enthusiastic amateur astronomer who uses a 0.46-m Dobsonian telescope for deep-sky observations. The name was suggested by the first discoverer.

Professor SCHMIEDMAYER, in Wien geboren, wird hier als Quantenphysiker und als begeisterter Amateurastronom geehrt.

(18398) **Bregenz**
E: 1992 Sep. 23 by F. Börngen at Tautenburg.
B: MPC 42366, 2001 Mar. 09.
Bregenz, capital of the Austrian province of Vorarlberg, is situated on the east shore of the Lake Constance and is known for the festivals on its famous seaside stage. The region was already settled in the Bronze Age. Later, the Romans established the mercantile town of Brigantium, and there was also an Alemannian colony.

Die Landeshauptstadt von Vorarlberg wird hier vom Österreich- und Musikliebhaber Dr. Börngen gebührend geehrt.

(19129) **Loos**
E: 1988 Jan. 10 by A. Mrkos at Klet'.
B: MPC 46012, 2002 Jun. 24.
Austrian architect Adolf Loos (1870–1933) was one of the pioneers of the functionalist style in Europe. Influenced by the rationalist architecture of the Chicago School, he rejected the Art Nouveau style. Loos' radical style influenced architects in Germany, Austria and Bohemia.

Adolf Loos wurde in Brünn geboren und starb in Kalksburg bei Wien. Zahlreiche Werke schuf er in Wien, war aber auch im Ausland tätig, wobei er bei aller Anerkennung auch Widerspruch erfuhr. – Die Frau des Grazer Astronomie-Professors Dr. Oskar Mathias (1900–1969), eine geborene Ernestine Loos (1911–1999), war eine Verwandte dieses Architekten.

(19612) **Noordung**
E: 1999 July 17 at the Črni Vrh Observatory.
B: MPC 59386, 2007 Apr. 03.
At the end of 1928 Herman Potočnik (1892–1929) published under the pseudonym Hermann Noordung a visionary work entitled *Das Problem der Befahrung des Weltraums – der Raketen-Motor*. In this work he was the first to detail a technical description of a space station in a geostationary orbit and its applications.

Herman Potočnik, Sohn slowenischer Eltern, geboren in Pola und aufgewachsen in Marburg, studierte nach dem Ersten Weltkrieg an der TU in Wien. Sein großes Interesse für Raketen- und Raumfahrt-technik resultierte in einem unter dem Pseudonym Hermann Noordung veröffentlichten Buch über Raumstationen und geosta-tionäre Satelliten. Er gilt als Pionier und Visionär der modernen Raumfahrt.

(19914) **Klagenfurt**
    E: 1973 Oct. 27 by F. Börngen at Tautenburg.
    B: MPC 42677, 2001 May 09.
Klagenfurt, the capital of the Austrian province of Kärnten, is situated on the eastern shore of the Wörther Lake in the greatest intramountainous basin of the Eastern Alps. This cultural center and tourist resort was first documented in 1195.

Klagenfurt am Wörthersee ist die größte Stadt des Bundeslandes Kärnten sowie die sechstgrößte Österreichs.

(21075) **Heussinger**
    E: 1991 Sep. 12 by L. D. Schmadel and F. Börngen at Tautenburg.
    B: L. D. Schmadel, MPC 56961, 2006 June 13.
Adalbert Heussinger (b. 1923) is a member of the Catholic order of Minorites. He was for 40 years confessor in St. Peter's Cathedral at the Vatican, where he absolved thousands of penitents of their sins. He is recognized as a distinguished scholar of theology and philosophy. The name was suggested by the first discoverer.

DDr. Pater Adalbert Heussinger lebt heute im Minoritenkonvent Graz und agiert weiter als treuer Seelsorger.

(21076) **Kokoschka**
    E: 1991 Sep. 12 by F. Börngen and L. D. Schmadel at Tautenburg.
    B: MPC 42678, 2001 May 09.
The Austrian expressionist painter, graphic artist and writer Oskar Kokoschka (1886–1980) was known for his portraits, landscapes and views of famous towns, produced in a monumental manner and expressive colors. He emigrated in the 1930s and from 1953 lived in Switzerland.

Kokoschka war geboren in Pöchlarn (Niederösterreich). Er kam in seiner Malerei vom Jugendstil zum Expressionismus, wirkte zunächst in Wien und ging nach einer kurzen intensiven Freundschaft mit Alma Mahler nach Dresden. Nach vielen Reisen von Europa bis zum Orient kam er nach Prag und floh im Jahre 1938 nach England. Er verstarb 1980 in der Schweiz.
    R: BdÖ 248.

(21109) **Sünkel**
    E: 1992 Sep. 04 by L. D. Schmadel and F. Börngen at Tautenburg.
    B: L. D. Schmadel, MPC 56961, 2006 June 13.

HANS SÜNKEL (b. 1948) is professor of mathematical geodesy and geo-informatics at the Technical University of Graz. A well-known expert on the shape of the earth, he currently serves as the university's rector. The name was suggested by the first discoverer.

Prof. SÜNKEL ist auch Direktor der Abteilung für Satellitengeodäsie des Institutes für Weltraumforschung der ÖAW in Graz. Derzeit (2008) ist er Rektor der Technischen Universität Graz. Er ist seit 1992 k.M. und seit 1998 w.M. der ÖAW.

(21564) **Widmanstätten**
E: 1998 Aug. 26 by E. W. ELST at the European Southern Observatory.
B: MPC 59386, 2007 Apr. 03.

Count ALOIS VON WIDMANSTÄTTEN (1754–1849) was an Austrian chemist who discovered typical patterns by acid etching iron meteorites. These unique Widmanstätten patterns can be used to determine if a piece of iron is in fact a meteorite.

Der bedeutende Meteoritenforscher ist in Graz geboren und in Wien verstorben.

(22322) **Bodensee**
E: 1991 Sep. 13 by F. BÖRNGEN and L. D. SCHMADEL at Tautenburg.
B: MPC 43048, 2001 July 05.

Bodensee, Constance Lake, or Swabian Sea, is a border sea of the Alps, adjoining Germany, Switzerland and Austria. The river Rhine flows through it. It is an old cultivated and cultural landscape. The name was suggested by the first discoverer.

(22618) **Silva Nortica**
E: 1998 May 28 by M. TICHÝ at Klet'.
B: MPC 60730, 2007 Sep. 26.

Silva Nortica is a historical name of the region covering the territory on the borders of South Bohemia and Lower Austria, well known for its pleasant landscape and historical sights. Since 2002 many forms of cross-border cooperation have taken place there.

Die Euroregion Silva Nortica (Nordwald) ist eine 2002 gegründete Organisation zur Förderung der grenzüberschreitenden Zusammenarbeit der Gemeinden Süd-Böhmens und des niederösterreichischen Waldviertels.

(24712) **Boltzmann**
E: 1991 Sep. 12 by F. BÖRNGEN and L. D. SCHMADEL at Tautenburg.
B: F. BÖRNGEN, MPC 43195, 2001 Aug. 04.

The Austrian physicist LUDWIG BOLTZMANN (1844–1906) confirmed Maxwellian electrodynamics experimentally in 1872 and is one of the founders of the kinetic theory of gases. He explained thermodynamics in terms of statistical mechanics and black-body radiation. The name was suggested by the first discoverer.

LUDWIG BOLTZMANN wurde in Wien geboren und starb in Duino bei Triest durch Selbstmord nach schwerer Krankheit. Er hatte in Wien studiert und lehrte mehrfach an den Universitäten von Wien und Graz als Professor für Experimentelle und Theoretische Physik und Naturphilosophie. Sein bewegter Lebenslauf spiegelte sich auch in der Mitgliedschaft bei der ÖAW, wo er mehrfach w.M. wurde (dazwischen gab es Auslandsaufenthalte).

R: NR Alm. d. ÖAW 57, 307 (1907); BdÖ 45.

(24916) Stelzhamer

E: 1997 Mar. 07 by E. MEYER at the Davidschlag Observatory, Linz.

B: MPC 47301, 2003 Jan. 06.

FRANZ STELZHAMER (1802–1874), great Austrian poet and novelist, studied law, theology and painting, and worked as journalist, teacher and author. His *s'Hoamatgsang* is the anthem of the Austrian province Upper Austria.

STELZHAMER ist durch seine Landeshymne ein bekannter Oberösterreicher.

(26074) Carlwirtz

E: 1977 Oct. 08 by H.-E. SCHUSTER at the European Southern Observatory.

B: MPC 52769, 2004 Sep. 28.

The German astronomer CARL WILHELM WIRTZ (1875–1939) worked at the Strasbourg and Kiel observatories. An untiring observer noted also for his orbit computations for comet D/1766 G1, he was the first, in 1924, to show statistically the existence of a redshift-distance relationship for spiral nebulae.

CARL WIRTZ war 1898 kurzfristig an der Kuffner-Sternwarte (KP 12568) tätig.

(26355) Grueber

E: 1998 Dec. 23 by E. MEYER at Linz.

B: MPC 50464, 2004 Jan. 07.

JOHANNES GRUEBER (1623–1680) was a Jesuit priest, missionary, mathematician and astronomer at the Chinese imperial court from

1659 to 1661. He returned to Europe from China by the overland route and published the very first travelogue describing Tibet.

Der Linzer JOHANNES GRUEBER gehörte zu einer Gruppe von Jesuiten und Astronomen, die am Kaiserhof in Peking eine ersprießliche Tätigkeit entfalteten und der dort unter anderem auch Astronomie lehrte.

    R: BdÖ 159.

(29227) **Wegener**
    E: 1992 Feb. 29 by F. BÖRNGEN at Tautenburg.
    B: MPC 45750, 2002 May 26.

The German scientist ALFRED WEGENER (1880–1930) created the theory of continental drift, which explained even the climates of prehistoric times and was verified later by geotectonics. A participant in several polar expeditions, he perished on the Greenland ice sheet.

Er studierte von 1900 bis 1904 Physik, Meteorologie und Astronomie in Berlin, Heidelberg und Innsbruck. 1924 erhielt WEGENER einen ordentlichen Lehrstuhl für Meteorologie und Geophysik in Graz, wo er endlich eine gesicherte Position für sich und seine Familie fand. In einer wagemutigen Polarexpedition beendete er sein Leben. Sein Andenken lebt fort im Wegener-Zentrum in Graz. WEGENER war k.M. der ÖAW seit 1925.

    R: NR Alm. d. ÖAW 81, 321 (1931); BdÖ 579.

(29427) **Oswaldthomas**
    E: 1997 Mar. 07 by E. MEYER at Linz.
    B: MPC 2004 Jan. 07.

OSWALD THOMAS (1882–1963), founder of the Astronomical Bureau in Vienna and of the Astronomischer Verein, is well known for his work on meteors and for popularizing astronomy. He proposed the establishment of the "Sterngarten", now known as the Vienna Open Air Planetarium.

OSWALD THOMAS wurde in Kronstadt (Siebenbürgen) geboren und ist in Wien gestorben. Er kam 1913 nach Wien und war zunächst Gymnasialprofessor; später Honorarprofessor an der Universität Wien. Viele erinnern sich auch heute noch an ihn oder zumindest an seinen Namen und die von ihm vertretene sehr wirksame Vortragstätigkeit.

    R: BdÖ 542, Biogr.: Der Sternenbote 25, 106 (1982).

(30788) **Angekauffmann**
    E: 1988 Sep. 08 by F. BÖRNGEN at Tautenburg.
    B: MPC 45237, 2002 Mar. 28.

The Swiss ANGELICA KAUFFMANN (1741–1807) painted pictures of mythological, allegoric and religious themes and became renowned for pleasing classicist portraits. She was a celebrated member of several European academies.

A. KAUFFMANN war die in Chur geborene Tochter des Malers JOSEPH KAUFFMANN, der aus dem Bregenzer Wald stammte. Ihr Bild zierte lange Zeit Österreichs 1000-Schilling-Banknote.

R: BdÖ 237.

(30844) **Hukeller**

E: 1991 May 17 by C. S. and E. M. SHOEMAKER at Palomar.

B: G. and D. HEINLEIN, MPC 57951, 2006 Nov. 09.

HANS-ULRICH KELLER (b. 1943) is a professor of astronomy at the University of Stuttgart. Since 1976 he has directed Stuttgart's Zeiss Planetarium and the Weizheim Observatory. He is also editor of the astronomical almanac – *Himmelsjahr*. The name was suggested by G. and D. HEINLEIN.

HANS-ULRICH KELLER studierte in Wien und ging, nachdem er einige Zeit bei ZEISS in Oberkochen und am Planetarium Bochum gearbeitet hatte, als Planetariumsleiter nach Stuttgart, wo er eine ungemein reichhaltige Tätigkeit entwickelte. Er hatte mehrfach Funktionen im Vorstand der Astronomischen Gesellschaft inne und ist seit 1997 Honorarprofessor für Astronomie in Stuttgart.

(30933) **Grillparzer**

E: 1993 Oct. 17 by F. BÖRNGEN at Tautenburg.

B: MPC 45345, 2002 Apr. 27.

The dramatic poet and playwright FRANZ GRILLPARZER (1791–1872) was a celebrated figure in Austrian literature, a lifelong loyal to the Habsburg dynasty and a friend of BEETHOVEN. Best known among his works are the historical tragedy *King Ottocar, His Rise and Fall* and the lyric tragedy *The Waves of Sea and Love*.

FRANZ GRILLPARZER gehört zu den größten Dichtern Österreichs, der in Wien geboren und gestorben ist. Er hinterließ ein reichhaltiges literarisches Werk, das ebenso wie seine Persönlichkeit in vielen Schriften gewürdigt worden ist. Grillparzer wurde 1847 zum wirklichen Mitglied der ÖAW gewählt.

R: Zahlreiche Nachrufe und Berichte über Feiern, Preise u. a. NR Alm. d. ÖAW 22, 211 (1872); BdÖ 155/157.

(36672) **Sidi**

E: 2000 Aug. 21 by the Lowell Observatory Near-Earth Object Search at the Anderson Mesa Station.

B: H. RAAB, MPC 51189, 2004 Mar. 06.

Foundling SIDONIE ADLERSBURG (1933–1943) grew up with her foster parents JOSEFA and HANS BREIRATHER in Sierning, Austria, until she was deported to Auschwitz, where she soon perished. She is memorialized in the novel *Abschied von Sidonie* by ERICH HACKL. The name was suggested by H. RAAB.

Sierning ist der Hauptort des Mühlviertels im Bezirk Steyr (OÖ).

(43955) **Fixmüller**

E: 1997 Feb. 06 by E. MEYER and E. OBERMAIR at the Davidschlag Observatory, Linz.

B: MPC 53471, 2005 Jan. 25.

PLACIDUS FIXMÜLLER (1721–1791), director of the astronomical observatory of the abbey in Kremsmünster, Upper Austria, worked on the orbit of Uranus and calculated a precise value for the solar parallax from observations.

Pater PLACIDUS FIXLMILLNER (verschiedene Schreibweisen) war an der Entwicklung der Astronomie in Kremsmünster intensiv beteiligt („Mathematischer Turm", das erste „Hochhaus" Europas). Er war vielseitig begabt und lehrte von 1746 bis 1787 als Professor für Kirchenrecht. Als erster Direktor der Sternwarte Kremsmünster (KP 6457) war er auch international sehr bekannt.

R: Zahlreiche Nachrufe und Biographien in den Akten von Kremsmünster; u. a. BdÖ 115.

(44613) **Rudolf**

E: 1999 Sep. 08 by P. PRAVEC and P. KUŠNIRÁK at Ondřejov.

B: MPC 50253, 2003 Nov. 09.

RUDOLF II. (1552–1612), of the Habsburg dynasty, was a Czech and Hungarian king and Roman emperor. He created an important center for arts and sciences in Prague, assembled a great artistic collection and employed TYCHO BRAHE and JOHANNES KEPLER as his court astronomers.

Der ausführlichen Biographie RUDOLFS II. ist wenig hinzuzufügen. Eines seiner großen Verdienste war, dass er die Astronomen TYCHO BRAHE (KP 1677) und JOHANNES KEPLER (KP 1134) an seinen Hof nach Prag verpflichtete.

R: BdÖ 441/44.

(48681) **Zeilinger**
E: 1996 Jan. 21 by E. MEYER and E. OBERMAIR at the Davidschlag Observatory, Linz.
B: MPC 53955, 2005 Apr. 07.

ANTON ZEILINGER (b. 1945) is a much-honored professor of experimental physics in Innsbruck and Vienna, decorated by the "pour le mérite" for sciences and arts. Well known for his contributions to quantum physics, in 1997 he succeeded in the first teleportation of information on quantum level.

ANTON ZEILINGER ist Mitglied zahlreicher Akademien und Träger verschiedenster Preise. Er wurde 1994 zum k.M. und 1998 zum w.M. der ÖAW gewählt und bekleidet dort wichtige Funktionen.

(48801) **Penninger**
E: 1997 Oct. 22 by E. MEYER at the Davidschlag Observatory.
B: MPC 57424, 2006 Aug. 09.

JOSEF PENNINGER (b. 1964), director of the Institute of Molecular Biotechnology of the Austrian Academy of Sciences, has been honored by various university chairs and numerous awards. He was chosen by the magazine *Esquire* as one of the ten most interesting people of the year 2000.

JOSEF PENNINGER leitet als wissenschaftlicher Direktor das 1999 gegründete Institut IMBA (Institut für Molekulare Biotechnologie GmbH). Er ist seit 2002 k.M., seit 2007 w.M. der ÖAW.

(49109) **Agnesraab**
E: 1998 Sep. 18 by R. LINDERHOLM at Lime Creek.
B: MPC 50465, 2004 Jan. 07.

Austrian amateur astronomer AGNES RAAB (b. 1969) is a long-time member of the Linzer Astronomische Gemeinschaft. The first prediscovery image of this minor planet was found on a plate exposed on her eighth birthday.

AGNES RAAB und ihr Gatte HERBERT RAAB (KP 3184) sind wohlbekannte Astronomen in Oberösterreich, deren Arbeiten über die reine Amateurtätigkeit weit hinausreichen.

(58191) **Dolomiten**
E: 1991 Dec. 28 by F. BÖRNGEN at Tautenburg.
B: MPC 50254, 2003 Nov. 09.

The Dolomiten is a mountain group with characteristic rocks formed of dolomitic limestone found in the Italian Alps. The highest point is the Marmolada (3342 m). The range takes its name from the French geologist DOLOMIEU (1750–1801), after whom the mineral dolomite was named.

(58499) **Stüber**
Dicovered 1996 Nov. 03 by E. MEYER and E. OBERMAIR at the Davidschlag Observatory, Linz.
B: MPC 59924, 2007 Jun. 01.
EBERHARD STÜBER (b. 1927) is director of the natural science museum "Haus der Natur" in Salzburg. Under his direction a space hall was established, so far the only permanent space exhibition in Austria. In 1988, STÜBER set up the Salzburg Public Observatory on Voggenberg as an outpost of the museum.

Die Planetenbenennung ehrt die oben angeführte wichtige Tätigkeit in der Volksbildung durch E. STÜBER in Salzburg.

(59001) **Senftenberg**
E: 1998 Sep. 26 by J. TICHÁ and M. TICHÝ at Klet'.
B: MPC 54567, 2005 July 21.
Senftenberg (now Žamberk) is where two comets were discovered by THEODOR BRORSEN in 1851. It is a pleasant market-town located at the foot of the Orlické mountains in eastern Bohemia. It is the birthplace of PROKOP DIVIŠ, astronomer AUGUST SEYDLER and surgeon EDUARD ALBERT.

Senftenberg ist eine liebliche Kleinstadt in Ostböhmen. – Siehe auch TH. BRORSEN (KP 3979), der an der dortigen Sternwarte seine Beobachtungen anstellte.

(65675) **Mohr-Gruber**
E: 1989 Jan. 11 by F. BÖRNGEN at Tautenburg.
B: MPC 51190, 2004 Mar. 06.
Curate JOSEPH MOHR (1792–1848) and his organist FRANZ XAVER GRUBER (1789–1863) are the creators of the carol "Stille Nacht, heilige Nacht" ("Silent Night! Holy Night!"), which was first heard on Christmas Eve in 1818 in Oberndorf near Salzburg.

Der Siegeszug des Liedes „Stille Nacht", das im Salzburger Land entstanden ist und über die ganze Welt verbreitet wurde, findet nun auch in der Planetenbenennung seinen Niederschlag.
R: BdÖ 351 (Mohr), 158 (Gruber). Der Entdecker F. BÖRNGEN wurde für die Benennung durch das Land Salzburg ausgezeichnet.

(69275) **Wiesenthal**
E: 1989 Nov. 28 by F. BÖRNGEN at Tautenburg.

B: MPC 52326, 2004 July 13.

SIMON WIESENTHAL (b. 1908) survived the Nazi camps of World War II. After the war, he courageously gathered data on the perpetrators of the Holocaust. He wrote several books, including *I Hunted Eichmann* and *The Murderers Among Us*.

SIMON WIESENTHAL, geboren in Galizien, der in zwölf verschiedenen Konzentrationslagern seine ganze Familie verloren hatte, deckte in seinem Dokumentationsarchiv viele Verbrechen gegen die Juden auf und konnte so zur Verurteilung der Schuldigen beitragen. Gestorben ist er hochbetagt im Jahre 2005 in Wien (begraben in Israel).

R: BdÖ 692.

(73700) von Kues

E: 1991 Oct. 05 by F. BÖRNGEN and L. D. SCHMADEL at Tautenburg.
B: MPC 53177, 2004 Nov. 26.

NIKOLAUS VON KUES (NICOLAUS CUSANUS, 1401–1464), born near Trier, was a theologian, mathematician, scholar, experimental scientist and influential philosopher. He stressed the incomplete nature of man's knowledge of God and of the universe. His paper "*Perfectio mathematica*" (1458) anticipates infinitesimal methods.

NIKOLAUS VON KUES war in Cues an der Mosel geboren und starb in Todi in Italien. Er wurde 1448 Kardinal und 1450 Bischof der Südtiroler Diözese Brixen. Er galt als Reformer und Freiheitskämpfer.

R: BdÖ 372.

(78391) Michaeljäger

E: 2002 Aug. 08 by S. F. HÖNIG at Palomar.
B: MPC 51982, 2004 May 04.

MICHAEL JÄGER (b. 1958) is one of the most prolific and recognized comet astrophotographers. Within the last decades he imaged around 300 different comets and discovered comet P/1998 U3 as well as fragment D of comet 141P/Machholz. In 2002 he assisted the discoverer with confirmation of comet C/2002 O4 (Hönig).

Herr JÄGER ist ein bekannter Journalist und, wie gesagt, ein überaus eifriger Kometenjäger, von dessen Aufnahmen bereits eine Broschüre erstellt wurde.

R: Mehrere Artikel im Sternenboten und zahlreiche Fotodokumentationen.

(85199) Habsburg

E: 1991 Oct. 03 by F. BÖRNGEN and L. D. SCHMADEL at Tautenburg.
B: F. BÖRNGEN, MPC 54177, 2004 May 04.

Habsburg or Habichtsburg ("hawk's castle") is a ruin in the Swiss canton of Aargau. It is the ancestral seat of the European HABSBURG dynasty, which reigned for 1000 years. Its power culminated with emperor KARL V. (1500–1558). The name was suggested by the first discoverer.

Die Habsburger sind ein europäisches Adelsgeschlecht, dessen Name sich von der Habsburg im Aargau (Schweiz), ihrer Stammburg, herleitet. Mitglieder der Habsburger-Dynastie herrschten jahrhundertelang über Österreich, Böhmen und Ungarn. Von 1438 bis 1740 gehörten alle Kaiser des Heiligen Römischen Reiches dem Haus HABSBURG an. Im 16. und 17. Jahrhundert herrschten sie auch über die Königreiche Spanien und Portugal und deren überseeische Besitzungen in Amerika, Afrika und Asien. Nach dem Tod des letzten männlichen Habsburgers, Kaiser KARL VI., trat die von dessen Tochter MARIA THERESIA mit FRANZ I. STEPHAN VON LOTHRINGEN begründete Dynastie HABSBURG-LOTHRINGEN ihre Nachfolge an und stellte von 1765 bis zum Untergang des Heiligen Römischen Reiches 1806 erneut die Kaiser; der letzte römisch-deutsche Kaiser, FRANZ II., begründete 1804 das erbliche Kaisertum Österreich, welches bis 1918 bestand. Nebenlinien der Dynastie HABSBURG-LOTHRINGEN regierten in der Toskana, Modena und Parma sowie kurzzeitig in Mexiko. Umgangssprachlich wird auch die noch heute bestehende Dynastie HABSBURG-LOTHRINGEN als „Habsburg" bezeichnet (aus Wikipedia).

(85317) Lehár
E: 1995 Jan. 30 by F. BÖRNGEN at Tautenburg.
B: F. BÖRNGEN, MPC 54177, 2004 May 04.
Composer FRANZ LEHÁR (1870–1948) created a new style of Viennese operetta. In 1905, he achieved worldwide success with *The Merry Widow*, *The Land of Smiles*, and other operettas followed. Several of his works were filmed.

FRANZ LEHÁR, der in Komorn (Ungarn) geboren wurde und in Bad Ischl starb, war zunächst Militärkapellmeister. Später widmete er sich ganz der Operette. Er hinterließ ein vielfältiges Werk von Märschen, Walzern, Liedern, Operetten und Opernmusik.
R: BdÖ 282.

(85389) Rosenauer
E: 1996 Aug. 22 by J. TICHÁ and M. TICHÝ at Klet'.
B: MPC 53472, 2005 Jan. 25.
JOSEF ROSENAUER (1735–1804) was the designer and master-builder of the Schwarzenbergs' Canal for floating timber from the Šumava

mountains to Vienna. Finished in 1793, this waterway connected the Vltava and the Danube, two rivers that flow into different seas.

Seine Karriere begann der in Kalsching Geborene als Forstadjunkt im Jahre 1759. Nach dem Studium in Wien erwarb er 1771 den Titel eines Grafeningenieurs, wurde beeideter Landvermesser und im Jahre 1791 Direktor des Schwarzenbergischen Wassertransportes. Er starb in Böhmisch-Krumau.

(85411) **Paulflora**

E: 1996 Nov. 03 by E. MEYER and E. OBERMAIR at the Davidschlag Observatory.

B: MPC 60301, 2007 July 30.

PAUL FLORA (b. 1922), caricaturist, graphic artist and illustrator, was born in South Tyrol but has lived in Innsbruck, North Tyrol, since his early years. His first book, FLORA's *Fauna*, was published in 1953. His ironic and sarcastic drawings, sketched in a distinctive, unique style, have gained international recognition.

Prof. PAUL FLORA, geboren 1922 in Glurns/Südtirol, ist ein freischaffender Zeichner, Karikaturist, Grafiker und Illustrator, der jetzt in Tirol lebt.

R: BdÖ 116.

(99201) **Sattler**

E: 2001 Apr. 25 by P. G. COMBA at Prescott.

B: H. WINDOLF, MPC 56615, 2006 Apr. 13.

BIRGIT I. SATTLER (b. 1969) is a member of the Department of Limnology and Zoology at the University of Innsbruck. She was a member of two Antarctica expeditions sponsored by the National Science Foundation and the Planetary Studies Foundation, respectively. The name was suggested by H. WINDOLF.

BIRGIT SATTLER wurde in Schwaz in Tirol geboren. Sie ist eine hochdekorierte Biologin und führend in ihrem Fachgebiet tätig.

(99861) **Tscharnuter**

E: 2002 July 29 by S. F. HÖNIG on NEAT images taken at Palomar.

B: MPC 54830, 2005 Sep. 18.

WERNER M. TSCHARNUTER (b. 1945) has made major contributions to the fields of star formation, protoplanetary disks, stellar dynamics and Saturn's rings. He also has an interest in celestial mechanics, particularly with regard to the evolution of the Koronis family, to which this minor planet probably belongs.

WERNER TSCHARNUTER, geboren in Mitterberg (bei Gröbming, Steiermark), studierte in Wien Astronomie, dissertierte bei Prof. FERRARI und wurde *sub auspiciis praesidentis* zum Dr. phil. promoviert. Er habilitierte sich in Göttingen und war dann Professor in Wien und Vorstand der Universitäts-Sternwarte, darauf wissenschaftlicher Mitarbeiter am MPI Garching und ging 1981 an das Institut für Theoretische Astrophysik nach Heidelberg, wo er jetzt noch tätig ist.

(117156) **Altschwendt**
    E: 2004 Aug. 23 by W. RIES at Altschwendt.
    B: MPC 55989, 2006 Feb. 19.
Altschwendt is a rural village in the north-western part of Austria. The rural environment with little light pollution provides exceptional conditions for astronomical observation. This is the first numbered minor planet discovered at the Altschwendt Observatory.

Altschwendt liegt bei Schärding (Oberösterreich) im nördlichen Innviertel, südwestlich von Peuerbach und beherbergt eine aufstrebende Privatsternwarte.

(128586) **Jeremias**
    E: 2004 Aug. 16 by W. RIES at Altschwendt.
    B: MPC 58598, 2007 Jan. 06.
JEREMIAS RIES (b. 2006) is the godchild of the discoverer. Jeremias is the German form of Jeremiah.

(144333) **Marcinkiewicz**
    E: 2004 Feb. 20 by W. RIES at Altschwendt.
    B: MPC 60732, 2007 Sep. 26.
EKHARD MARCINKIEWICZ (b. 1928) is a well-known amateur astronomer in Austria. He documented the sun accurately and consistently in thousands of drawings over many decades and has played an important role in Austrian solar research.

(154865) **Stefanheutz**
    E: 2004 Sep. 09 by W. RIES at Altschwendt.
    B: MPC 60302, 2007 July 30.
STEFAN HEUTZ (b. 1980) is a German jurist and amateur astrophotographer. His pictures and articles are published in books, magazines and in the internet. The naming of this minor planet is in appreciation of his efforts to encourage and promote astronomy.

HEUTZ ist ein Freund und Kooperationspartner (Amateurastronom) des Entdeckers.

3. Alphabetischer Index der Planetennamen mit ihren zugehörigen Nummern

Schrödinger	13092	Struveana	768	von Suttner	12799
Schwarzschilda	837	Stüber	58499		
Seeligeria	892	Suess	12002	Wegener	29227
Semper	6353	Sünkel	21109	Weizsäcker	13531
Senftenberg	59001	Švejk	7896	Werfel	12244
Shapleya	1123			Widmanstätten	21564
Sidi	36672	Trenker	17489	Wiesenthal	69275
Silva Nortica	22618	Tscharnuter	99861	Wilkens	1688
Šindel	3847	Tycho Brahe	1677	Wolfgangpauli	13093
Spencer Jones	3282				
Stefanheutz	154865	Vietoris	6966	Zachia	999
Stefanzweig	17597	Vogelweide	9910	Zauberflöte	14877
Stelzhamer	24916	von Kues	73700	Zeilinger	48681
Stifter	7127	von Matt	9816	Zuckmayer	8058

4. Alphabetischer Index und Nummern der in der ersten Arbeit I (SCHNELL und HAUPT, 1995) behandelten Planetennamen

Jena	526	Mozartia	1034	Schoenberg	4527
Jubilatrix	652	Murray	941	Schrutka	2665
Justitia	269			Schubert	3917
		Natalie	448	Schulhof	2384
Kafka	3412	Nausikaa	192	Scott	876
Kallisto	204	Neally	903	Scylla	155
Kálmán	4992	Neunkirchen	4216	Semmelweis	4170
Kärnten	6451	Noemi	703	Silesia	257
Katharina	320	Nora	783	Silvretta	1317
Kepler	1034			Sita	244
Kleopatra	216	Oceana	224	Siwa	140
Klotilde	583	Olga	304	Sophia	251
Konkolya	1445	Oppavia	255	Stampfer	3440
Kovacia	867	Oppolzer	1492	Steiermark	6482
Kriemhild	242	Oskar	750	Stephanie	220
				Strauss	4559
Lacrimosa	208	Palisana	914	Stumpff	3105
Lameia	248	Pannonia	1444	Subamara	964
Leonisis	728	Patroklus	617	Susi	933
Leontina	844	Paulina	278		
Libera	771	Pawona	1152	Tamara	326
Lindemannia	828	Penelope	201	Theresia	295
Linzia	1469	Perseverantia	975	Thora	299
Liszt	3910	Petzval	3716	Thule	279
Lorenz	3861	Philagoria	274	Thusnelda	219
Lucia	222	Philia	280	Tinette	687
Lucretia	281	Picka	803	Tirol	6439
Ludovica	292	Polana	142		
		Probitas	902	Valda	262
		Protogeneia	147	Valentina	447
Mahler	4406	Purgathofer	5341	Vindobona	231
Marmulla	711			Vorarlberg	6332
Martha	205	Rakos	4108		
Mathilde	253	Rosa	223	Walpurga	256
Mayrhofer	1690	Rosenkavalier	5039	Webern	4529
Medea	212	Roxane	317	Weringia	226
Melanie	688	Russia	232	Widorn	3721
Meliböa	137			Wolfiana	827
Mendel	3313	Salzburg	6442		
Misa	569	Sapientia	275	Xantippe	156
Montefiore	782	Schober	2871	Zita	689

5. Anhang: Wie und wann erhalten Kleinplaneten ihre Namen?

(1) Die Namensgebung eines einzelnen KP steht am Ende eines langen Prozesses, der sich über viele Jahre, manchmal Jahrzehnte, hinziehen kann.

(2) Es beginnt mit der Entdeckung eines KP, der nicht mit einem schon bekannten Objekt identifiziert werden kann. Wenn Beobachtungen von mindestens zwei Nächten dem MPC

gemeldet werden, vergibt dieses eine so genannte Provisorische Bezeichnung.

(3) Es ist dann immer noch möglich, dass der KP mit einem schon früher entdeckten Objekt identifiziert wird. In diesem Fall gehen die Entdeckerrechte an diesen ersten Beobachter über – er/sie erhält die so genannte „principal designation".

(4) Werden keine älteren Beobachtungen gefunden, wird der KP aber über längere Zeit – Wochen oder Monate – weiter beobachtet, bestehen gute Chancen, ihn bei der nächsten Opposition wieder zu finden.

(5) Je nach Typ oder Art der Bahn sind zwischen zwei bis vier Oppositionen notwendig, um einem KP eine definitive (laufende) Nummer zu geben.

(6) Der Entdecker eines nummerierten KP ist identisch mit dem Inhaber der „principal designation".

(7) Diesem steht das Privileg zu, einen Namen für die Entdeckung vorzuschlagen, und es gilt für einen Zeitraum von zehn Jahren seit der Nummerierung des KP.

(8) Der Entdecker schreibt eine kurze Begründung (Citation), die den gewählten Namen erklärt. Unter anderem ist zu beachten, dass

– Namen von Personen und Ereignissen, die hauptsächlich wegen ihrer militärischen oder politischen Tätigkeiten bekannt geworden sind, nur dann akzeptiert werden, wenn 100 Jahre seit dem Tod einer Person vergangen sind bzw. seit das Ereignis stattgefunden hat.
– Namen von Haustieren sind nicht erwünscht.
– Namen mit ausschließlich oder überwiegend kommerziellen Absichten sind nicht erlaubt.

(9) Ein 15-köpfiges Komitee (Committee for Small-Body Nomenclature) der Internationalen Astronomischen Union (IAU), bestehend aus Berufsastronomen aus der ganzen Welt, entscheidet über die vorgeschlagenen Namen.

(10) Angenommene Namen werden als offiziell erklärt, sobald sie in den Minor Planet Circulars veröffentlicht worden sind. Dies geschieht normalerweise jeden Monat (um die Zeit des Vollmondes herum).

Ausführlichere Angaben findet man auf der Webpage des MPC unter

http://cfa-www.harvard.edu/iau/info/HowNamed.html

Danksagung

Für die Unterstützung unserer Arbeit danken wir Herrn F. BÖRNGEN, Frau M. G. FIRNEIS, Herrn L. D. SCHMADEL, Frau A. SCHNELL und Frau I. VAN HOUTEN-GROENEVELD sehr herzlich.

Literatur

[1] HAUPT, H., HOLL, P. (2000) Datenbank Österreichischer Astronomen (1330–2000). Verlag der Österreichischen Akademie der Wissenschaften, Wien

[2] HERGET, P. (1955) The Names of Minor Planets. Cincinatti Observatory, pp. 1–38

[3] HERGET, P. (1968) The Names of Minor Planets. Cincinatti Observatory, pp. 39–138

[4] KLEINDEL, W., VEIGL, H. (1987) Das Große Buch der Österreicher (immer nur mit der zugehörigen Seitenzahl zitiert). Kremayr & Scheriau, Wien

[5] PALUZÍE-BORELL, A. (1963) The Names of Minor Planets and Their Meanings. Jean Meeus, Kesselberg Sterrenwach Kessel-LO, Belgien

[6] SCHMADEL, L. D. (2003) Dictionary of Minor Planets. Names, 5th ed. + CD. Springer, Berlin Heidelberg New York

[7] SCHNELL, A., HAUPT, H. (1995) Kleine Planeten, deren Namen einen Österreichbezug aufweisen (I). Sitzungsber. Öst. Akad. Wiss. Wien, math.-nat. Kl., Abt. II **204**: 185–257

[8] WEISS, W. W., VYORAL-TSCHAPKA, M. (1982) Die Kuffner-Sternwarte in Wien-Ottakring. Sitzungsber. Öst. Akad. Wiss. Wien, math.-nat. Kl., Abt. II **191**: 615

[9] WURZBACH, C. VON (1856–1891) Biographisches Lexikon des Kaisertums Österreich (60 Bände). Wien

Anschrift der Verfasser: Em. Prof. Dr. Hermann Haupt, Waltendorfer Hauptstraße 141, 8042 Graz, Austria. E-Mail: hermann.haupt@uni-graz.at; PD DDr. Gerhard Hahn, Institut für Planetenforschung, Deutsches Zentrum für Luft- und Raumfahrt (DLR), Rutherfordstraße 2, 12489 Berlin, Deutschland. E-Mail: gerhard.hahn@dlr.de.

Sitzungsber. Abt. II (2007) 216: 127–134

Sitzungsberichte

Mathematisch-naturwissenschaftliche Klasse Abt. II
Mathematische, Physikalische und Technische Wissenschaften

© Österreichische Akademie der Wissenschaften 2008
Printed in Austria

Zur Kombinationswirkung von Schall und Erschütterungen

Von

Peter Steinhauser

(Vorgelegt in der Sitzung der math.-nat. Klasse am 13. Dezember 2007
durch das k. M. I. Peter Steinhauser)

Zusammenfassung

Den aktuellen Diskussionen im Bereich des Lärmschutzes folgend wird der Stand der Forschung zur Kombinationswirkung von Lärm und Erschütterungen auf den Menschen zusammenfassend dargestellt. Insbesondere die unterschiedlichen Wahrnehmungsformen – Erschütterungen über den Körper, Schall durch das Ohr – erschweren experimentelle Untersuchungen. Gesicherte Ergebnisse sind, dass der Mensch bei energetisch schwachen Immissionen (Verkehr) zwischen Hören und Spüren nicht unterscheiden kann. Eine Kombinationswirkung ist erst ab W_m-bewerteten Schwingbeschleunigungen von $30\,mm/s^2$ und A-bewerteten Schallpegeln von 45 dB nachgewiesen. Ab etwa $100\,mm/s^2$ bzw. 65 dB werden Verdeckungseffekte für die jeweils andere Wahrnehmungsform wirksam. Untersuchungen zur Quantifizierung der Kombinationswirkung im betroffenen Immissionsband stehen noch aus. Im Vergleich zur Immissionswirkung von Schall und Erschütterungen alleine dürfte die Kombinationswirkung aber nur untergeordnete Bedeutung besitzen.

1. Einleitung

Schwingungen, die sich durch feste Körper zum Aufenthaltsbereich von Menschen hin ausbreiten, können von diesen einerseits direkt als Erschütterungen und andererseits als von der Oberfläche einer Bauwerksstruktur (i. a. von einer Decke oder Wand) abgestrahlter sekundärer Luftschall wahrgenommen werden. Dabei werden die Erschütterungen, da es ein eigenständiges Erschütterungsorgan nicht gibt, taktil durch

Tabelle 1. Menschliche Wahrnehmung und Empfindung von direktem Luftschall, Erschütterungen und Sekundärschall

Immissionsform	Wahrnehmung	Empfindung
direkter Luftschall (Lärm)	auditiv durch das Ohr	Fernreiz, Ortung, Flucht möglich
Erschütterung	taktil durch den ganzen Körper	Nahreiz, Abwehrreaktion unmöglich
Sekundärschall	auditiv durch das Ohr	Nahreiz, Abwehrreaktion unmöglich

den ganzen Körper und der Sekundärschall – ebenso wie der direkte Luftschall – auditiv durch das Ohr aufgenommen. Diese beiden Wahrnehmungsformen bilden somit zwei Aspekte der gesamtheitlichen Erschütterungseinwirkung und somit eine kombinierte Exposition.

Über die Wahrnehmung hinaus ist auch die Empfindung des Menschen für diese Immissionsformen zu berücksichtigen. Dabei ergeben sich Empfindungsunterschiede bezüglich des direkten Luftschalls einerseits und der Erschütterungen sowie des Sekundärschalls andererseits.

Beim direkten Luftschall, der seiner großen Reichweite wegen als Fernreiz bezeichnet werden kann, sind im Freien die Ortung der Schallquelle und in Gebäuden Schutzmaßnahmen wie Fensterschließen möglich: Es wird dabei der Urinstinkt der Fluchtmöglichkeit angesprochen. Erschütterungen und der Sekundärschall stellen hingegen Nahreize dar, denen gegenüber man das Gefühl des Ausgeliefertseins erleben kann. Klassische Lärmschutzmaßnahmen wie Lärmschutzfenster sind zwecklos. Bei den Erschütterungen kommt noch hinzu, dass das biometrische Schwingungsverhalten einzelner Körperteile (Resonanzphänomene) Wahrnehmung und Empfindung unter Umständen beeinflussen können.

In Tab. 1 werden diese unterschiedlichen Wahrnehmungs- und Empfindungsformen schematisch dargestellt [1].

Dass es sich dabei um die Empfindungskategorien Lästigkeit und Störung handelt, kommt sprachlich bereits in den Bezeichnungen Lärm und Erschütterungen zum Ausdruck, die für positive Empfindungen wie „Musik hören" oder „in den Schlaf wiegen" nicht verwendet werden.

2. Hörbarkeit und Fühlbarkeit

Um diese Kombinationswirkung objektiv beurteilen zu können, ist es wie bei allen Immissionsformen zuerst erforderlich, sie quantitativ zu erfassen, d. h. operational durch Messverfahren, Zu- oder Abschläge und/oder Bewertungsfaktoren zu bestimmen. Dies wird durch die oben dargestellte unterschiedliche Form der Wahrnehmung erschwert, weshalb derzeit erst Aussagen zu Teilaspekten möglich sind. Die allgemeine Interaktion von Lärm und Erschütterungen wird für die Wahrnehmungsempfindung von verschiedenen Autoren unterschiedlich beantwortet, wie aus verschiedenen Übersichtsartikeln übereinstimmend hervorgeht [1–3].

Da die verschiedenen Untersuchungen naturgemäß unterschiedliche Teilaspekte betreffen, ergibt sich außerdem ein lückenhaftes Bild der Kombinationswirkung. Bezüglich der Erschütterungswahrnehmung ergibt sich durch Schallpegel $L_A \geq 64\,$dB nach MELONI [4] ein Verdeckungseffekt. STAHL et al. [5] untersuchen die Frage, ob ein Zusammenhang zwischen der Kombinationswahrnehmung von Schall und Erschütterungen, der daraus abgeleiteten Distanzwahrnehmung und der erlebten Lästigkeit besteht. Nach SCHUST et al. [6] ergibt sich bei der Kombinationswirkung ein zunehmender Einfluss der Tonhaltigkeit auf die Beurteilung von Lautheit und Belästigung.

Zumindest bei energetisch schwachen Immissionen (z. B. Verkehr) können in der Praxis belastete Personen oft nicht unterscheiden, ob sie etwas gehört oder am Körper gespürt haben. Die Begriffe „Hören" und „Spüren" werden von den Betroffenen oft unterscheidungslos verwendet. Dazu kommt, dass die Wahrnehmbarkeit von Erschütterungen sehr stark situationsabhängig ist. Nur bei gespannter Aufmerksamkeit wird eine Erschütterung der W_m-bewerteten Schwingbeschleunigung von $3{,}6\,$mm/s^2 nach ON-ISO 2631-1 tatsächlich wahrgenommen werden können. Jede Ablenkung (Zuhören, Lesen etc.) vermindert die Wahrnehmbarkeit bereits beträchtlich und eigene Aktivitäten (Reden etc.) noch mehr. Dementsprechend kann die Fühlschwelle nicht als Schwellwert, sondern nur als Schwellenband angegeben werden. Die obere Bandbegrenzung liegt etwa bei einer bewerteten Schwingbeschleunigung von $15\,$mm/s^2 [7], wie dies bei verschiedenen Untersuchungen immer wieder erkennbar ist [8].

3. Schienenverkehrsimmissionen

Hinsichtlich der Schienenverkehrsimmissionen liegt eine umfassende Studie von SAID et al. [9] vor, in deren Rahmen neben dem Auflösungsvermögen der menschlichen Wahrnehmung für die Stärke von Erschütterungen auch die Kombinationswirkung von Lärm und Erschütterungen auf den Menschen untersucht worden ist. Der Versuchsablauf wurde dabei nach den Methoden der Psychophysik so angelegt, dass neben der Sinnesempfindlichkeit der Probanden auch ihre Reaktionsbereitschaft erfasst werden konnte, da bei Diskriminationsaufgaben „entdeckungsfreudige" Personen mögliche Differenzen schneller melden als „zurückhaltende" Personen, die sich ihrer Sache zuerst sicher sein wollen.

Es wurde die Unterscheidung von insgesamt vier Erschütterungsintensitätsstufen ($K_{B,\max} = 0{,}2/0{,}4/0{,}8/1{,}6$ entsprechend $a_{W_{m,\max}} = 7{,}1\,\mathrm{mm/s^2}$, $14\,\mathrm{mm/s^2}$, $29\,\mathrm{mm/s^2}$ und $57\,\mathrm{mm/s^2}$ bei drei A-bewerteten Vorbeifahrtspegeln (30, 45 und 55 dB) geprüft. Die Bandbreite der Stärke beider Immissionsformen umfasst somit einen Bereich, der im Umweltschutz häufig zu nachdrücklichen Klagen führt. Die Versuche ergaben

- A-bewertete Vorbeifahrtspegel von 30 dB sind ohne Einfluss auf die Unterscheidung der Erschütterungsintensität.
- A-bewertete Vorbeifahrtspegel von 45 dB beeinflussen die Unterscheidung ab einer W_m-bewerteten Schwingbeschleunigung von $29\,\mathrm{mm/s^2}$ aufwärts.
- A-Bewertete Vorbeifahrtspegel von 55 dB reduzieren die Unterscheidungsfähigkeit um mehr als 20 %.
- Trotz verhältnismäßig hoher Intensitäten beider Immissionsformen sind die Kombinationseffekte eher tendenziell als signifikant nachweisbar gewesen.

In einer Untersuchung über die Lärm-Erschütterungs-Kombinationswirkung von Shinkansen-Hochgeschwindigkeitszügen von YOKOSHIMA und TAMURA [10] wird ebenfalls berichtet, dass die Shinkansen-Züge negativer zu bewerten sind als der sonstige Schienenverkehr, wo die Immissionen geringere Stärke besitzen. Da in dieser Studie einerseits unbewertete Schwingbeschleunigungsdaten für die Erschütterungsimmissionen verwendet wurden und andererseits die zugrunde liegende Bausubstanz der japanischen Studie (87 % Holzhäuser) für europäische Verhältnisse untypisch ist, sind die sonstigen Ergebnisse der Untersuchung nicht mit den europäischen Arbeiten vergleichbar.

Eine weitere Untersuchung zur Kombinationswirkung von Schall und Erschütterungen des Schienenverkehrs haben HOWARTH und GRIFFIN [11] mit dem Ziel durchgeführt festzustellen, welche der beiden Immissionsformen stärker belästigend wirkt und deshalb vordringlich zu bekämpfen ist. Es wurden Zugsvorbeifahrten von 24 Sekunden Dauer mit jeweils 6 verschiedenen Vorbeifahrtspegeln des Schalls bzw. 6 unterschiedlichen Erschütterungsintensitäten hinsichtlich der Belastigungseinstufung durch die Probanden untersucht. Die

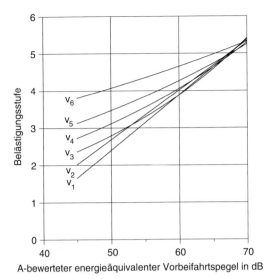

Abb. 1. Einfluss der Erschütterungsintensität auf die Gesamtbelästigung kombinierter Schall- und Erschütterungsimmissionen für W_m-bewertete Schwingbeschleunigungen (nach [11]) (v_1: 23 mm/s^2; v_2: 32 mm/s^2; v_3: 45 mm/s^2; v_4: 65 mm/s^2; v_5: 90 mm/s^2; v_6: 130 mm/s^2

untersuchte Bandbreite sowohl der Erschütterungsintensität als auch des Schallpegels liegt oberhalb jener der Studie von SAID. Da dabei die Erschütterungsintensität in der gemäß ON-ISO 2631-1 nicht anzuwendenden Kenngröße VDV (Vibration Dose Value) angegeben ist, muss diese zur besseren Vergleichbarkeit in die W_m-bewertete Schwingbeschleunigung transformiert werden. Dies kann, da die beiden Einheiten in keinem formelmäßigen Zusammenhang stehen, nur näherungsweise nach einer Schätzformel [3] erfolgen. Das Ergebnis ist in den Abb. 1 und 2 wiedergegeben.

Nach Abb. 1 ergibt sich für A-bewertete Vorbeifahrtspegel von 45 dB eine deutliche Erhöhung der Belästigung durch gleichzeitige Erschütterungsimmissionen von 65 mm/s^2 aufwärts. Diese zusätzliche Belästigung nimmt mit anwachsendem Schallpegel ab, und spätestens ab 65 dB beeinflussen die Erschütterungsimmissionen den Grad der Belästigung nicht mehr (Verdeckungseffekt). Bei niedrigen Erschütterungsintensitäten bis etwa 45 mm/s^2 ist kein statistisch signifikanter Einfluss auf die Belästigung durch die Schallimmissionen erkennbar, wie die sich mehrfach schneidenden Regressionskurven erkennen lassen.

Analog ergibt sich nach Abb. 2 für die W_m-bewertete Schwingbeschleunigung bei etwa 25 mm/s^2 eine deutliche Erhöhung der Belästi-

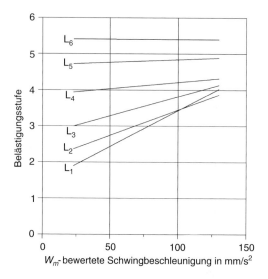

Abb. 2. Einfluss des Schallpegels auf die Gesamtbelästigung kombinierter Schall-
und Erschütterungsimmissionen für A-bewertete Vorbeifahrtspegel (nach [11])
(L_1: 45 dB; L_2: 50 dB; L_3: 55 dB; L_4: 60 dB; L_5: 65 dB; L_6: 70 dB)

gung durch gleichzeitige Schallimmissionen. Auch hier verringert
sich die zusätzliche Belästigung mit anwachsender Erschütterungsintensität, wobei die Erschütterungen ab einem A-bewerteten
Schallpegel von etwa 60–65 dB für Belästigung insignifikant werden
und somit den in Abb. 1 erkennbaren Verdeckungseffekt bestätigen.
Auch im Fall der Erschütterungsimmissionen ergeben niedrige
Vorbeifahrtsschallpegel (45–50 dB) keinen signifikanten Einfluss auf
die Belästigung, wie die sich schneidenden, eng beieinander liegenden
Regressionskurven belegen.

Aus diesen Untersuchungen zusammen ergibt sich somit ein
komplexes Bild der Kombinationswirkung. Im Bereich von Erschütterungsintensitäten unterhalb des Schwellenbands der Fühlbarkeit kann es
definitionsgemäß keine Kombinationswirkung geben. Erschütterungen
geringer, jedoch spürbarer Intensität können von Schallimmissionen
praktisch nicht unterschieden werden, sodass keine Wechselwirkung
beobachtet werden kann [1]. Erst bei mittleren Erschütterungsinten-
sitäten (ca. $a_{W_m} = 30\,\mathrm{mm/s^2}$) bzw. Schallpegeln (ca. $L_{A_{eq}} = 45\,\mathrm{dB}$)
fächert die Belästigungswirkung in Abhängigkeit von beiden Im-
missionsgrößen auf, bis ab A-bewerteten Schallpegeln von etwa
60–65 dB ein Verdeckungseffekt des Schalls wirksam wird.

Das interessante Ergebnis dieser Untersuchung ist es, dass die
Versuchspersonen der Erschütterungsreduktion bei eher schwachen

Erschütterungsimmissionen und der Lärmreduktion bei hohen Lärmimmissionen den Vorzug geben.

4. Schlussfolgerungen

Hinsichtlich der Kombinationswirkung von Erschütterungen und Sekundärschall können die Ergebnisse der hier diskutierten Untersuchungen folgendermaßen zusammengefasst werden.

1. Bei schwachen Immissionen (z. B. Verkehr) kann der Mensch zwischen hörbaren Schall- und fühlbaren Erschütterungsimmissionen nicht unterscheiden, weshalb eine spezielle Kombinationswirkung nicht fassbar ist.
2. Eine Wechselwirkung zwischen Erschütterungs- und Schallimmissionen ist erst ab W_m-bewerteten Schwingbeschleunigungen von 30 mm/s^2 aufwärts zu erwarten, wenn die A-bewerteten Vorbeifahrtsschallpegel auf 45 dB und darüber anwachsen.
3. Wenn eine der beiden Immissionsgrößen auf niedrigem Niveau bleibt, wie dies beim Sekundärschall praktisch immer zutrifft, dann ergibt sich keine signifikante Kombinationswirkung.
4. Die Immissionsschutzkriterien von ÖNORM S 9012 für Erschütterungs- und Sekundärschallimmissionen des Schienenverkehrs liegen unter den oben angegebenen Werten. Es ist daher nicht erforderlich, eine Kombinationswirkung beider Immissionsformen zusätzlich zu berücksichtigen.

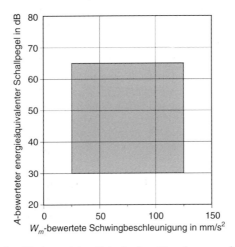

Abb. 3. Bereich der W_m-bewerteten Schwingbeschleunigung und der A-bewerteten Schallpegel, für den weitere qualitative und quantitative Untersuchungen der Kombinationswirkung erforderlich sind

5. Hohe Schallpegel (ab etwa 60–65 dB) bewirken einen Verdeckungseffekt für Erschütterungen.
6. Die Quantifizierung der Kombinationswirkung innerhalb des in Abb. 3 angegebenen Immissionsbereichs ist noch durchzuführen.

Insgesamt gilt immer noch die Feststellung von GRIFFIN [11], die sinngemäß lautet: „Die potenziellen Wechselwirkungen der Reaktion auf Lärm und Erschütterungen sind komplex. Der Nachweis solcher Effekte ist schwach und teilweise widersprüchlich." Beispielsweise sind weitere Untersuchungen zur Kombinationswirkung für niedrige und mittlere Immissionsstärken (siehe Abb. 3) von Schall und Erschütterungen erforderlich.

Literatur

[1] MELONI, T. (2006) Bericht der Fachgruppe zur Verordnung über den Schutz vor Erschütterungen. Bundesamt für Umwelt BAFU, Abt. Lärmbekämpfung, Bern
[2] HAIDER, M., KOLLER, M., STIDL, H. G. (1992) Qualitätskriterien für Schienenverkehrslärm und Erschütterungen bei Vollbahnen – Kombinationswirkungen von Lärm und Erschütterungen. Forschungsarbeiten aus dem Verkehrswesen, Band 36/1, Wien
[3] GRIFFIN, M. J. (1990) Handbook of Human Vibrations. Academic Press, San Diego
[4] MELONI, T. (1991) Wahrnehmung und Empfindung von komplexen, kombinierten Belastungen durch Vibrationen und Schall. Dissertation, ETH Zürich
[5] STAHL, E., MELONI, T., KRUEGER, H. (1997) Distanzwahrnehmung von Umweltsituationen. Zs. f. Lärmbekämpfung 44/1
[6] SCHUST, M., SEIDEL, H., BLÜTHNER, R. (1998) Wirkung von Lärm unterschiedlicher Tonhaltigkeit nach DIN 45681 in Kombination mit Schwingungen. Zs. f. Lärmbekämpfung 45/4
[7] ON-ISO 2631-1 (2007) Mechanische Schwingungen und Stöße – Bewertung der Auswirkung von Ganzkörperschwingungen auf den Menschen, Teil 1, Anhang C
[8] BERGER, P., LANG, J., ÖSTERREICHER, M., STEINHAUSER, P. (2005) Wirksamkeit der Schutzmaßnahmen gegen U-Bahn-Immissionen für den Wiener Musikverein. Zement und Beton 2/05
[9] SAID, A., FLEISCHER, D., KILCHER, H., GRÜTZ, H. P. (2001) Zur Bewertung von Erschütterungsimmissionen aus dem Schienenverkehr. Zs. f. Lärmbekämpfung 48/6
[10] YOKOSHIMA, S., TAMURA, A. (2005) Combined annoyance due to the Shinkansen railway noise and vibration. Internoise Congress, Rio de Janeiro
[11] HOWARTH, H. V. C., GRIFFIN, M. J. (1990) The relative importance of noise and vibration from railways. Applied Ergonomics 21/2

Anschrift des Verfassers: Univ.-Prof. Dr. Peter Steinhauser, Delugstraße 8, 1190 Wien, Österreich. E-Mail: Peter.Steinhauser@univie.ac.at.